# Mossy Memoir of a Rolling Stone

# MEMOIRS AND OCCASIONAL PAPERS
## Association for Diplomatic Studies and Training

In 2003, the Association for Diplomatic Studies and Training (ADST) created the Memoirs and Occasional Papers Series to preserve firsthand accounts and other informed observations on foreign affairs for scholars, journalists, and the general public. Sponsoring publication of the series is one of numerous ways in which ADST, a nonprofit organization founded in 1986, seeks to promote understanding of American diplomacy and those who conduct it. Together with the Foreign Affairs Oral History program and ADST's support for the training of foreign affairs personnel at the State Department's Foreign Service Institute, these efforts constitute the Association's fundamental purposes.

# Mossy Memoir of a Rolling Stone

## Thompson Buchanan

ASSOCIATION FOR DIPLOMATIC STUDIES AND TRAINING
MEMOIRS AND OCCASIONAL PAPERS SERIES

NEW ACADEMIA PUBLISHING · VELLUM

Washington, DC

Printed in the United States of America

Library of Congress Control Number: 2011935911
ISBN 978-0-9836899-4-2 paperback (alk. paper)

VELLUM   An imprint of New Academia Publishing

NAP   NEW ACADEMIA PUBLISHING

New Academia Publishing
PO Box 27420, Washington, DC 20038-7420
info@newacademia.com - www.newacademia.com

# Contents

To my dear family Nancy, Barbara, and Campbell, who may regret that they encouraged me to set out on this egotistical journey

and

to all my Foreign Service companions, remembered and forgotten, whose help, advice, companionship, humor, and patience made rolling along with the Foreign Service so rewarding

# Prologue

Retired Foreign Service officers are often told in real or feigned admiration, "What a fascinating life. You must write it up." But it is usually the family who are the most insistent that we resurrect the past. The moss of time may have affected the accuracy of these memoirs of a rolling stone, but so what? There are few still alive to contradict us.

This is the story of the peripatetic life of a successful but not particularly distinguished State Department official caught up in the *Sturm und Drang* of the Cold War and its affiliated conflicts. This public servant spent most of his thirty-three years as an analyst and Foreign Service officer. The reader who perseveres will, one hopes, be left with a better appreciation of life in the Foreign Service trenches and a capsule view of life overseas, ranging from Europe to many years involved with Russia and Central Africa, as well as some concluding thoughts on the art of diplomacy and life itself. The author hopes readers will enjoy some of the vignettes along this memory lane, including a nostalgic account of a house and friends in the French Provence. They may even learn that behind this seemingly exotic façade of Foreign Service life are real folks suffering all the complex emotions and travail of the average American family, along with a few extra strains that are inherent in Foreign Service life.

# Acknowledgments

This memoir benefited from the kind advice of friends and family, with a special thanks to the transcriber of this manuscript, Cheryl Battles, for her patience and suggestions. Inevitably, ranging through so much time and place, it gives a superficial view of some important events. But these are the things that stick in the mind about events long ago, some of them as vivid as though they happened yesterday.

My omission of names of colleagues and of foreign diplomats who played an important role in the events that I describe reflects only an editorial preference for brevity—and certainly no reflection of my evaluation of them as friends and colleagues.

A final word of thanks is due Margery Thompson, publishing director of the Association for Diplomatic Studies and Training, who undertook the chore of turning this manuscript into something that may appeal to a circle of readers beyond the Foreign Service.

Mossy Memoir of a Rolling Stone

# 1

# Those Formative Years

Socially, I fitted the pattern of the traditional pre–Rogers Act Foreign Service, but the resemblance stopped there. For I was born in 1924 not on the East Coast but in Beverly Hills, California, where my father, Thompson Rodes Buchanan, a former playwright, was a longtime editor of Goldwyn Pictures. My mother, Katherine Winterbotham, a society lady from Chicago, had lived in Berlin before the war, studying singing and reportedly living with a German doctor. She spoke nostalgically of her time in Berlin, riding in the Tiergarten and having the Kaiser tip his hat to her, and being asked to sing in a German opera. This worldly lady finally could not tolerate my father's philandering with the ladies he hired for his plays and divorced him in 1927, when I was only three years old. My mother reportedly said that her greatest mistake was marrying my father instead of continuing her singing in Germany. I rather appreciated her lack of judgment.

The first of my journeys as a rolling stone was by train to a Chicago very different from the city of modern skyscrapers and beautiful gardens of today. I grew up on Oak Street near Michigan Avenue, which in those days consisted of great blocks of cement that oozed tar in the torrid summer heat and formed ice mountains in winter. One day I watched a father and son swept off a mount of ice by a wave.

Chicago in those days was the Wild West of the Middle West. This is not surprising if we remember that Fort Dearborn was built to protect the city from the Indians in 1802, and the last Indian massacre was in 1812. In my childhood, the savages were the gangs that fought over the profits from prohibition. I recall the excitement over the Saint Valentine's Day Massacre and the explosion of the house across the street when something went wrong with their production of homemade gin. One night I was awakened by the sound of male voices outside my mother's bedroom, where a man in police uniform and two in civilian clothes were explaining to my five-foot-two imperious mother that they had penetrated our chained front door in pursuit of a robber. When she told the family bootlegger, who made his usual delivery the following day, he turned white and fled. Perhaps he ended up as one of the many blocks of human cement dumped into Lake Michigan in those days of vicious gang warfare.

This was the height of the Great Depression, with men begging at the back door for a bowl of soup. The young woman who helped my mother was trying to support a crippled husband who had fallen into a pit of molten metal. She used to take me to the park, where she pointed out the people trying to keep warm by stuffing their clothes with newspapers. As a result, I shared with many Depression era youths that guilty sense of privilege that attracted them to socialism, and those of lesser judgment to communism.

My mother's family, the Winterbothams, relatively old Chicagoans, were very socially conscious, concerned with genealogy and the Blue Book. My later years at an upper class school in England only added to my revolt against what I came to see as the injustice of a class society. The Winterbothams, sometimes spelled Winterbottom, originated in Northern England, where they wintered cattle from the Highlands. As a "master cloth maker," John Winterbotham was forbidden to leave England in order to preserve England's monopoly in the field of cloth. He and his wife were smuggled out of England with the help of a General Humphrey on General Washington's staff and settled in Connecticut in 1811. It was their son, John Humphrey Winterbotham, who moved to Michigan City, Indiana, where he started a successful cooperage business, using convict labor. After the Civil War, in which he was badly wounded, Colonel Winterbotham served two terms "with distinction" in the Indiana Senate. His son, Joseph Humphrey Winterbotham, was also a successful businessman. It was he who donated the Winterbotham collection of French impressionist paintings to the Art Institute in Chicago.

My grandmother used to attend the symphony in her electric car, spend the summer at her houses in Sugar Hill, New Hampshire, and on one occasion in Chicago had to be reminded by the doorman that she had forgotten to put on a skirt. My Uncle John became the *paterfamilias*, managing the family monies in the trust company he steered successfully through the Depression but mismanaged in the postwar era, when stocks replaced mortgages as the focus of investment. My uncle's greatest distinction, perhaps, was the number of Martinis he could consume throughout the day. When he died, my cousin David Baldwin, a doctor, insisted on doing an autopsy. He was astounded to find that Uncle John had the liver of a thirty-year-old—apparently preserved in gin.

A memorable moment in the family life in Sugar Hill occurred when the butler appeared at my grandmother's with a trayful of cocktails and she told them she was fed up with their drinking behind her back. The family had always stopped at the farmhouse for a "wee dram" before climbing up the hill to Grandmother's.

It was my uncle who checked on the suitability of the various men anxious to marry my rather beautiful, socialite mother, among them a Viennese baron. My mother liked titles and the exotic. Scandal tore our family apart when my mother was introduced by a society lady in Chicago to Hindu Prince Jehan Kumar Warliker Seesodia, a graduate of Harrow and Cambridge and a London barrister on a lecture tour of America. It was headline news in the *Chicago Tribune* when they were married over strenuous opposition by both my uncle and my father. To intimidate my mother, my father enticed my sister Katherine, or Babs as she was called, who was six years older than I and close to her father, to spend the weekend with him. As a result we became separated, with my sister being brought up by her maiden aunt in Louisville, Kentucky, while I continued my travels, this time abroad, with mother, my new stepfather, and my French nurse Zelle, who did not approve of the marriage. We fled first to Toronto, Canada, where my father and uncle went to court to take custody of me on the grounds that my stepfather was not a Caucasian. Rather than risk an adverse decision, Zelle and I were put on one boat to France and Mother and Jehan took another. We settled in Cannes on the Riviera in that prewar Anglo-Saxon expatriate community that enjoyed, but was never really part of, France.

**Early Years Abroad**

My rolling stone existence in Europe got off to a disastrous start in a French *école primaire*. I was then given a tutor in French and boxing lessons and transferred to a Catholic school called Stanislas. To prepare this rather sickly boy for the rigors of English school life, I attended a Swedish exercise class on the beach in Cannes throughout the winter. Then we moved to Glion sur Montreux in Switzerland, where I endured daily cold baths prescribed for me by the Swiss doctor. As a day student at the local English Maccata school, I learned to skate, pick the glorious narcissi covering the mountainside, and not to argue with Hungarians who knew judo.

We then moved to London, where I was enrolled in a preparatory school, Stanmore Park, which was then in the country and later became the Balloon Barrage Headquarters of the RAF in World War II. It was expected at Stanmore that I would go to Harrow. At Stanmore, which later merged with Stratton Park, both schools of barely 50 students, it was no great achievement to be both a leading student and athlete, helped by a teacher, my own Mr. Chips, who took a personal interest in making me a good little Englishman.

With war looming on the horizon and my mother's wish to educate her son in America, we returned to New York in 1938. There, my mother and Jehan tried on a limited income to live on Park Avenue and Sutton Place in the style to which they had become accustomed in Europe. I followed in the family footsteps by attending Phillips Exeter Academy and Yale University. It was indeed a shock to transfer from the very ordered life of a tiny English school to a school like Exeter, with 600 boys and few rules except the requirement of academic excellence. My posh English accent did not help, but soccer learned in England smoothed the transition.

Outside of school activities, I was absorbed in the news about the Battle of Britain and decided at one point that I should leave for Canada to enlist in the RAF. It was then that sister Babs, who had come back into my life from Wellesley College, dissuaded me from this heroic gesture with an ease that I still find embarrassing.

**Life in Kentucky**

Babs never ceased envying the exotic life she presumed I had led in Europe, where I often felt alienated and alone, and I in turn envied the American roots she put down in Louisville, home to the Buchanan and Thompson families. According to my sister's Colonial Dames papers, both the Buchanans and the Thompsons traced their American roots back to the seventeenth and eighteenth centuries, including a member of the Maryland House of Burgesses. My Old Kentucky Home was also part of my heritage. Judge Rowan, a relative by marriage, built what was officially known as Federal Hill at Bardstown, but it was his cousin Stephen Foster who immortalized the house with his song. Judge Rowan's daughter Anne married Joseph Rhodes Buchanan, a well-known scientist, in 1841; hence our connection to Bardstown. Before donating a family portrait of Judge Rowan's son, Atkinson Hill Rowan, to Bardstown, we had it evaluated and found that it was painted in Spain, presumably when Atkinson was on a mission for President Andrew Jackson and before dying from cholera.

Living with her Aunt Mildred, Babs attended high school in Louisville and taught a class at the University of Louisville. Following the death of her first husband, Colonel Gilbert Elliot, she returned to Louisville, where she married Charles Birnsteel, the widowed husband of an old friend. In so doing, she literally returned to her roots, for she is buried in the family cemetery at Kane Station in the center of Louisville, along with the father I never knew, Aunt Mildred, and some slaves. My wife and I and our children will be joining her in due course in this place with gravestones going back to the seventeenth century.

A certain sense of adventure apparently ran in both my family and that of my Philadelphia wife, Nancy Shaw, as reflected in our missionary ancestors. Aunt Mildred, who brought up my sister, had worked for and then directed St. Hilda's School for Girls in Wuhan, China, from 1916 to 1927, when civil war forced them to evacuate the school. Years later, when I met with the vice principal in Wuhan of what has become a high school for fifteen hundred boys and girls, he showed me a history in Chinese of the former girl's school.

Nancy's Grandfather Shaw, for his part, sailed with his new bride on their honeymoon out to his new Presbyterian mission in Harbin, China, in the nineteenth century. Some honeymoon, on a rickety ship on a stormy sea!

# 2

# The World War II Years

In 1942 I graduated Cum Laude from Exeter, one of thousands whose college careers were cut short by World War II. In my brief eighteen months as a civilian at Yale, I lived the life of a typical college student, including being thrown out of the Elliot House at Harvard for disorderly conduct, to the amusement of my dean at Yale. As at Exeter, I was on the varsity soccer, squash, and tennis teams at Yale. I still wince at the memory of the sanctimonious lecture I gave on the evils of fraternities to a fellow soccer player who kindly asked me to join his fraternity. A German major, I took an eclectic mix of courses, from German poetry to Greek mythology, along with international relations under a brilliant British professor—all while waiting to join the Navy V-12 College Training Program.

My V-12 program began inauspiciously with a Captain's Mast, because I had gone AWOL when leave was canceled, too late for me to call off a date in New York. I suspect it was my girlfriend's uncle, who was an admiral, and not my eloquence that saved me from being sent to boot camp. Standing guard in the cold in my gob's uniform outside Saybrook College dampened my emotional ties to Yale.

In October 1944 I graduated from the U.S. Naval Reserve Midshipman School on a cockroach-infested ship in the Hudson River near Columbia University. My shipmates included some Southern boys who boasted about lynching "niggers" and an Italian who kindly warned me that he was a Golden Gloves champion before we got into a fistfight. With my lack of any scientific background or interest, my marks were mediocre.

As someone resistant to naval discipline, I chose as my onward assignment a landing ship tank (LST), among the smallest vessels afloat with the least requirement for naval protocol. But on the day of graduation, a Commander Hindmarsh asked that anyone who knew a foreign language should raise his hand. The next thing I knew I was traveling to Colorado to join that "very secret" Naval Language School in Boulder. There I was told that I was to learn, not Japanese as I expected, but Russian.

Boulder changed the whole course of my life, for not only did it provide me with Russian, which became the bread and butter of my professional career, but also brought me my wife of now sixty-six years, Nancy Shaw of Philadelphia, a transferee to the University of Colorado from Sarah Lawrence, who wanted the experience of a coed university. Nancy was one of three girls living in what one could only call a delightful "man trap," for the three officers dating the three girls all married them. For their honeymoon, I encouraged my future best man Palmer Smith and his Dotty to go to Jackson Hole, Wyoming, where I had hiked in 1939, never realizing how momentous a decision that proved to be. I was faced with the problem of winning a girl who was a proficient skier who was being courted by members of the 10th Mountain Division, not to mention some of my own language school colleagues. The first time that I flew in a plane Nancy was the pilot. I was not a naturally competitive person, but I somehow won out in the competition for Nancy's affection and she accepted my proposal of marriage.

Before telling our parents of our plans, I introduced Nan to my sister and to the Winterbothams in Chicago, where she was declared the best-looking girl in the family. But my parents were distressed. For my mother, her "little boy" was too young to marry. My stepfather, in a revealing statement, warned me against marrying someone richer than myself, saying that it can destroy love. My mother-in-law Bobo was upset, I suspect, that instead of marrying another fox hunter from the Philadelphia Main Line, Nancy had chosen someone who must be a communist because he was studying Russian. But we persevered and were married at Nan's home on July 26, 1945, on her terrace overlooking the hills where she loved to ride.

From the bride's perspective in particular, our three-day honeymoon on Nantucket was inauspicious. We missed our train connection, spent our first night in a hotel of dubious repute in New Haven, and almost starved on a train with no dining car. And, more ominous than all, the groom inadvertently took his typewriter with him on his honeymoon.

## Short Naval Intelligence Career

We had been taught Russian to help hand over lend-lease vessels to the Russians in Alaska. With my charming teachers, Prince and Princess Mestchersky of Saint Petersburg, we discussed culture but nothing about the Navy. Most of us were sent to Washington and told to work with our seven-months Russian on making a Russian dictionary. One day I was summoned to a Navy captain who asked how I would feel about going out to become the simultaneous interpreter for General MacArthur. Knowing something about what is required of real interpreters, I showed a certain lack of enthusiasm and was then blackballed by the captain for more appropriate assignments. And so, as a German major at Yale, I arranged to get myself assigned to a naval intelligence unit in Bremen, Germany, in January 1946. En route, an admiral in Frankfurt accused me of being un-American because I was eating European style with my fork in my left hand.

Much of Bremen remained in rubble from the bombings, and life was hard for the Germans. So, naturally the German girls flocked around our naval unit seeking food and marriage. As a newly married man, I became the father confessor to some nice girls who feared they would be considered prostitutes. I shared the agony with one girl of waiting in vain to hear from one of my colleagues who had left, promising to return to marry her. My only local contact, the head of the German Rundfunk, a radio station, told me how even his close friends would hear no criticism of *Der Fuehrer*. Watching our naval unit doing target practice with 45- caliber revolvers, the bedraggled German soldiers returning home from the front must have wondered how they could have lost the war.

Aside from discussing the bad smell of fish on the Bremerhaven dock with a Soviet naval officer there to supervise the transfer of German vessels to the Soviet Union, I had little use for my Russian. That is, until my captain returned from a trip to the Soviet zone of

Germany with a KGB document so secret, he said, that "we would both be dead if the KGB knew that we had it." Some East German con man must have sold him what proved to be water-smudged reports by a low-level KGB informant on some sunken riverboat of what "Igor said to Ivanov." My captain would not believe me and insisted on sending the document urgently to Washington, Top Secret, where it probably sits today in a highly classified Navy vault.

A trip up to Berlin in my Jeep to try lining up a job with the military government after I left the Navy in May 1946 proved quite unsuccessful, and I got lost in the Soviet zone on my return, with Mongolian soldiers jumping into my Jeep at checkpoints to ask for cigarettes and Penicillin for their officers. Cigarettes were the local currency in occupied German, but I had no cigarettes left, having used them up buying paintings in the Rosen Gallerie on Kurfürstendam: four prints by George Grosz, two Barlach wood prints, a Max Beckman, and a Käthe Kollwitz, for cigarettes. The little lady chortling over the sensual doodles of lady's posteriors on the back of the Grosz prints turned out to be Grosz's aunt.

# 3

# Back to Civilian Life

## Wyoming

Liberated from the Navy, Nancy and I set out for Jackson Hole, Wyoming, with my sister and her husband, Colonel Gilbert Elliot, to spend the summer on the ranch that Nan had bought before our wedding. My sister finally realized that "cattle guards" were not men on horseback. While Nancy was a successful competitor at the Devon Horse Show in Philadelphia, she also had Western roots in Sheridan, Wyoming, used to ride herd, and dreamed of having her own ranch.

My future best man Palmer Smith, another willing victim of the "man trap," took his bride on their honeymoon to Jackson Hole. Knowing Nan's dream of owning a ranch, the Smith's ears perked up when they were told of a ranch whose owner Dave Edmiston was forced to sell but was "damned if I'm going to sell to dem Rockerfellers." For he was one of several small ranchers under pressure to sell to the Snake River Land Company, an anonymous front company set up by John D. Rockefeller to restore the whole valley to its pristine wilderness. Philadelphia ranching friends of Nan's mother reassured her that the 320-acre ranch was a great buy at $19 an acre!

The ranch became not only a fantastic dowry, which helped finance both the construction of our house in Washington and the purchase of a house in France, but also a wonderful, enriching educational experience for the whole family. My first summer I worked as a carpenter and wire-tier on an Oliver hay bailer, walking four miles to work along Fish Creek with my little lunch pail. To provide gravity-feed water to the cabin, I had to build a 6x5x4-foot settling basin up in the woods, where the concrete truck was saved from turning over by a large Douglas fir. To bury the pipe from a spring high up the mountainside, to protect it from rodents and moose hooves leading down through the thick woods to my settling basin, I borrowed a horse and plow from a friend in Idaho in exchange for three days of tossing around eighty-pound bales of hay. Our toilet was a three-hole outhouse, "Inspiration Point," looking out over the valley. Our hot water came from a water jacket in our old Homelite woodstove attached to a barrel, and our bath was a large round metal tub. On one occasion, after a bone-breaking day, I got stuck in the tub and Nan had to pry me out. There were plenty of dead trees to cut down for firewood. We had no electricity or telephone, relying for light on white gas or kerosene lamps with mantles and for water on a little creek that ran near the cabin. Our first year, when we had no car, we borrowed horses, which Nan would ride the six miles for groceries to Hungry Jack's grocery store in Wilson. We were told that we could have a young, big Grey if we could catch him. It took us two days to catch him, but after a month he followed us around like a dog.

All this changed in 1981, when I retired from the State Department. To enlarge our existing cabin, we bought and trucked a cabin 75 miles from the ranch of Struthers Burt, a writer of the West heavily involved with John Rockefeller in establishing the Grand Teton National Park. His wife was startled when one day, after her husband's death, a park ranger came and told her that the fifty-year lease agreed by her husband was up and she had a year in which to move out of the ranch. Her shock was our good fortune, enabling us to tie Mrs. Burt's mother's two-bedroom cabin on to our own three-room cabin. Only then did we become decadent, and after thirty-six years of roughing it we installed all the luxuries of home—electricity, telephone, bathroom, and TV.

Along with the hard labor, we had a wonderful life hiking to distant mountain lakes, fishing, camping out in the woods on cut pine boughs, on one occasion carrying our skis along with our children up to Snow Lake. We used to watch the excellent little rodeo at the bar in Wilson, listening to the yarns from the old timers.

Jackson in those days was a simple cow town, known for the crooked gamblers in the Cowboy and Silver Dollar bars until 1952, when the women of Jackson chased them out of town. We had one policeman, Woodie, who gave me my first driving license. He sat me on the wooden sidewalk, coached me with some tricky questions, and said: "I know you can drive. I saw you drive into town." The valley was also a very democratic community, with the Rockefellers and the Burts joining the cowboys and dudes in the bar at Moose.

In 1947, Nan and I had our first and only experience of serious mountain climbing, scaling the 13,800-foot Grand Teton with Glenn Exum — a beautiful hunk of man, who pioneered the Exum Trail and taught violin in Idaho in the winter. Having somehow descended 200 feet upside down for my only lesson in rappelling, held only by a rope behind my knee, I was understandably a bit nervous when it came to dropping into the void over a great boulder high on the mountain. Nan, who has vertigo, managed the climb by never taking her eyes off Glenn.

There was a certain romance in the shrill cry of the coyotes in our lower pasture, the scream of a snow rabbit they killed, and once, the eerie cry of a mountain lion stalking a baby moose and the bellowing of the cow moose. Fortunately most of the bears in our area were black bears, often yearlings kicked out of the nest after one year to make it on their own. One of our tenants found one of these hungry yearlings eating her brownies beside her as she cooked them in her kitchen. Another found a bear behind him in his pickup truck. A bear destroyed our icebox, the old-fashioned kind that we had to feed with blocks of ice from town.

The porcupines were our greatest nuisance, chewing the salty logs of our cabin in the early hours of the morning. Our golden retriever Nicholas never learned about porcupines. The first time we had to take him down to our neighbors who had one electric light so that I could extract some 150 quills with a pair of pliers.

On our last visit to Jackson, driving home at 10:30 at night, two huge black elk jumped over the hood of our car, one of them destroying our windshield. We were extremely lucky.

John Rockefeller's vision of the valley he loved and sought to protect has been destroyed. The valley today is not Millionaire but Billionaire Valley, with gated communities in some places where hay fields used to dot the valley. Former Vice President Cheney and

Secretary of State James Baker have built mansions in the valley. To be sure, life has become easier for the poor ranching families with the arrival of K-Mart and better schools. And there's an excellent annual music festival at Teton Village, attracting musicians from all over the United States. But a way of life that we loved is no more. It was therefore with sadness but no great trauma that we and our children, Barbara and Campbell, decided to sell what remained of Nan's original ranch. With the money they were able to buy several houses for rental income.

## Back to Academe

Picking up my story: Following my first summer in Jackson I returned to the East to complete my one remaining term at Yale. The student body, with many returning veterans, was noticeably more serious. We lived in a home for delinquent children, where Nancy worked eighty hours a week supervising girls almost her age and much older in experience. I studied, played soccer for Yale, and wondered what I was to do after graduation in January 1947.

I concluded that my parents were probably right in thinking that, with my background, the Foreign Service would be a logical career. Accordingly, I arranged an appointment in Washington with a Mr. Green in charge of the Foreign Service exam. He did not approve of my plan of first taking a graduate degree in Europe, apparently preferring to take in new FSOs when they were still young and malleable.

I persisted with my plan of getting a graduate degree in Europe, postponing any FSO exam. Traveling to Europe was an adventure. I had been accepted at the Institut des Hautes Etudes in Geneva, Switzerland. It was too late to apply to the Sorbonne. Our liner, the old *Kungsholm*, burned at the dock in New York, forcing us to take a Black Diamond Liberty ship for an unusually rough crossing. We had been told that our Dachshund Heidi would give birth at sea, but a leading specialist in New York had told Nan that she definitely was not pregnant. But Nan was indeed pregnant. It cost three-quarters of my $90 monthly stipend under the GI Bill of Rights to get our ancient wardroom trunks across ruined Antwerp to the train for Paris, where we almost starved.

In Geneva, we found a little pension in Le Petit Sacconet, where the weekly fare was pasta varied with potatoes. I assumed that the change in my wife's silhouette was a result of the diet and had her exercising on the floor. We found that it was cheaper to join a tennis club where we could take a shower than pay for a bath in the pension. Shocked to learn that we were going to be parents, we decided to make the most of our remaining time of liberty: skiing in Chamonix and squeezing in a tour of Italy and Southern France, revisiting sites from my childhood.

Almost broke, we got passage on a train for Geneva, so jammed that the only seat Nan could find was on the men's toilet. Under the circumstances, it was probably not surprising that our daughter Barbara was two months premature, following a difficult birth without anesthesia. The only incubator in Switzerland was in Zurich and was occupied, so Barbara was swaddled in cotton wadding surrounded by warm beer bottles. Our local angel, M'Liz Schwob, Californian wife of a leading Swiss businessman and mother of three, took over with clothes and good advice. For several weeks I would bicycle up the long hill to La Grangette Clinic with milk pumped by Nan. To celebrate when Barbara finally came home, M'Liz opened a bottle of Mum champagne, and jealous Heidi retreated under the bed and bit off her dewclaw.

### Job-hunting in America

With a daughter, I had to think about getting a job and forgetting about a European degree. Nan flew back to her home in Philadelphia, Barbara in one basket and Heidi in the other, while I sailed home on a Liberty ship. My best offer in private enterprise was $40 a week for American Express in New York editing the president's correspondence. Armed with a letter from former secretary of labor, Francis Perkins, an old family friend, and the Homburg and gloves that my mother insisted I wear, I peddled my wares with no takers at the State Department.

Embarrassed at living for months at my mother-in-law's house, I was pleased to accept a job offer from Naval Intelligence at what was then the handsome salary of $5,000 a year—until I learned that I was to become a cryptanalyst working to break Soviet codes, not my cup of tea. Happily, Boris Klosson, head of the State Department's Foreign Political Branch, Division of Research on the Soviet Union (DRS), Office of Intelligence Research, the successor at the time to the World War II Office of Strategic Services (OSS), came to my rescue. Having also attended the Geneva Institute, he was inclined to overlook my homburg. Even though the job paid only $3,600 a year, I accepted it with alacrity in March 1948.

## Our Washington Home

The Washington we came to in 1948 bore little resemblance to the Washington of today. There were black families living on each side of our house at 2915 N Street in Georgetown, each with its own outhouse. One day when our wonderful black secretary Gloria had a birthday, I stupidly suggested that we go to Kitty and Al's across from the State Department after work and have a drink. Kitty was very apologetic but refused to serve Gloria, and so we adjourned to the nearby apartment of Ben Zook from Tennessee.

Almost every day as I walked down 23rd Street to the State Department, I followed Secretary of State Dean Acheson and Supreme Court Justice Felix Frankfurter, Mutt and Jeff—tall Dean with his cuffed pants two inches above his shoes, and Felix with his pant bottoms dragging on the ground, usually roaring with laughter. Security was not an issue.

In pursuing our cultural life, we were embarrassed to watch Margot Fonteyn and her fellow dancer Robert Helpmann falling on the stage of the infamous Constitution Hall of the Daughters of the American Revolution (which had snubbed Marion Anderson), because of the wax buildup on their shoes. Today, there are many venues for ballet in Washington, including a superb Kirov Academy of Ballet.

In 1950 we had moved into a basement apartment on Ordway Street, because we could not afford our monthly rent of $168 and had decided we needed to buy a house. Later that year our Swiss friend, Robert Schwob from Geneva, passed through Washington and complained that his wife, M'Liz, who had helped us with Barbara's birth, kept asking him to contact someone called Strasburg. We got Bob to call and it ended with the Strasburgs, who had children the same age as ours, persuading us to look at a piece of property, the original Fort Sumner overlooking the Potomac River on the Maryland side.

Looking at Little Falls below through the honeysuckle wilderness, listening to the muffled roar of the water, we were hooked. And so we forked out $8,000, a large sum at the time, to buy one acre from Chalmers Roberts, diplomatic correspondent for the *Washington Post*.

Since loans were hard to get during the Korean War, we reached into Nan's "dowry" of our Jackson Hole ranch and sold a large portion of the ranch to our neighbor, Stan Resor, former secretary of the army, for $78 an acre—a pittance compared to land values in the valley today. But this enabled us to build our Finnish-style modern house for $24,000 in 1951, before I joined the Foreign Service. We called our community of seven houses "The Hill." It was a wonderfully civilized family environment in which to bring up children.

# 4

# Early Days at State

The head of the Division of Research on the Soviet Union, Mose Harvey, looked like Lenin and was also an ideologue. I was initially intimidated by Mose, speaking at machine-gun speed with his deep Georgian accent, and by my colleagues. Several were ex-journalists, tapping expertly on their typewriters while I struggled with blocked keys in my effort to learn the touch system. Self-doubt was probably the cause of my sinus headaches, which drove me regularly to walk down to the Potomac River to get away from work. Annex 1, which was located where now there is an overpass above Virginia Avenue on 23$^{rd}$ Street, had no air- conditioning, so that our sweaty fingers would skid off the keys typing.

**Research on the Soviet Union**

My areas of analytic responsibility were Soviet "front" organizations, arms control, Soviet-UN relations, Soviets in Latin America, and Soviet-U.S. bilateral relations. I was essentially a Kremlinologist, drawing not always accurate conclusions from a careful perusal of the Soviet press and other limited sources of information. I caught Mose's eye with an analysis I made linking growing Soviet militancy in 1949 with Soviet expectations of a depression that would weaken America.

In a sense I joined Mose as the office ideologue, able to quote Stalin and Lenin at length. But the longer I worked on Soviet affairs, and particularly after interrogating defectors and service in Moscow, I realized that ideology often served mainly to justify what the leadership wanted to do but not to shape policy in the fundamentalist way that I originally believed.

I remain intrigued by a study I made in 1955 of the factors underlying the changes in the Soviet leadership at the time of Stalin's death. Looking over the theoretical press, I found what appeared to be an internal debate on whether Russia was still threatened by "capitalist encirclement" or whether "encirclement" no longer existed, with the implication that there could be some "withering away of the state" as postulated in communist theory. At the 19th Communist Party Congress in 1952 and early 1953, Stalin's presumed successor Georgy Malenkov attacked the "vulgarisers of Marxism" who allegedly called for "the weakening and dying off of the state in the atmosphere of capitalist encirclement."

The death of Stalin in March 1953 apparently averted another 1937-style "purge" of the top leadership. The reputed plans of Lavrentii Beria, minister of internal Affairs (MVD), to negotiate some deal on the neutralization of Germany that emerged after Stalin's death may have explained some curious articles in the Soviet press at the time. Intriguing attacks on "capitulationists" may have been directed against those officials who sought some détente with the West in an effort to block Western plans to build up Germany as a counterweight to Russia. Encouraged by Stalin to discuss an alternative German policy, Beria could have fallen into a Stalin trap and exposed himself as a pro-German "capitulationist."

With the death of Stalin, Mose Harvey at State and Charles "Chip" Bohlen in Moscow argued so bitterly over the meaning of the "collective leadership" that had emerged that serious analysis was stalled. We analysts silently disagreed with Mose's dogmatic insistence that nothing would change under communism. Churchill also urged Eisenhower to test and shape attitudes in post-Stalin Russia, but Eisenhower listened to the cautious "cold warriors" around him.

**Reporting from the Trenches**

During the Korean War, we worked a 24-hour watch shift in OIR. On one such occasion on the night shift, I noted that the Communist Party of Denmark was calling for peace negotiations in Korea. Given Moscow's tight control of its foreign communist parties, this aroused a great deal of interest and proved in fact an early indication that the Soviets and Koreans were prepared to blink.

It was on such slight variations from the norm that much of Kremlinology was based. But our predictions were sometimes dreadfully off. In preparing our monthly publication *Soviet Affairs*, our Korean analyst was faced with conflicting reports regarding a North Korean attack on South Korea, but weighing the evidence he concluded that the attack would not take place. That issue of *Soviet Affairs* was published after the North Korean invasion on June 21, 1956. A CIA friend, who has sadly passed on, bet me a keg of beer that the Korean invasion was the start of World War III.

It was my impression that OIR played a more important role in State in the early postwar years, in part because the Soviet desk in the European Bureau did not have the staff and expertise to produce the many documents required for the several four-power Council of Ministers conferences. The result is that we were often asked to produce documents for the desk at the last minute. It was in 1949 when I was asked to pull together for the following day a history of our relations with Berlin, about which I knew very little. After working all night, we staggered across the street to Kitty and Al's for a drink.

Shortly before I joined the Foreign Service in 1955, we were tasked by John Foster Dulles one Friday to produce by Sunday a report demonstrating that the Soviets would back down from their threats if met with firmness—a view I happened to share with the secretary. Dulles congratulated me and my deputy, Ben Zook, for the report, which was distributed to NATO.

Mose Harvey was proud of the team he had put together and was therefore furious when a number of us opted to integrate into the Foreign Service. My brilliant aide, Helmut Sonnenfeldt, remained a civil servant and went on to become adviser on Soviet policy in the department. Hal was much like Kissinger, both in background, as an immigrant from Germany, and in personality. When I suggested to him that he might achieve more by being less abrasive, he retorted that, unlike me, "born with a silver spoon in my mouth," he had had to fight his way up in life.

# 5

# Foreign Service Begins

Some of my FSO friends had urged me to join the Foreign Service. The Wriston Program, integrating the Civil Service and the Foreign Service, facilitated the decision that Nancy and I had already made on an anniversary stroll in the Virginia woods. And thus in August 1955 the Buchanan family of four (we now had a son named Campbell) arrived in a cold, rainy Frankfurt to my first atypical Foreign Service post. Even though the Montague Piggots, the consul general and his wife, were leaving on our first day, Mrs. Piggot insisted that Nancy call on her with all the protocol of the pre–Rogers Act generation—long gloves, hat, and visiting cards per the old Foreign Service. We were told how Mrs. Piggot had refused to allow any FSO wife to work and would tell them what they should wear to a reception.

Happily, a relaxed bachelor and future director general of the Foreign Service, John Burns, replaced Piggot. Despite the commiseration of German hands, we enjoyed Frankfurt on two counts—it was our first post, and my job was rather exotic.

## Interrogating Defectors

With our isolation from the Russian people, George Kennan had thought Soviet defectors could provide State Sovietologists with unique insight into Soviet attitudes on a full range of issues from politics to sex. A small unit attached to our consulate general was accordingly set up in the CIA Defector Reception Center.

We had access to the defectors only after they had passed a tough initial interrogation to establish their *bona fides*. It was a difficult but inevitably subjective task. In one case where a Jewish philosophy student from the University of Moscow was denied *bona fides* and dumped on the German economy, we suspected that the interrogators might have been prejudiced against someone who was both an intellectual and a Jew. Another sophisticated defector who had found employment with Radio Liberty broadcasting into the Soviet Union left in disgust over the ethnic squabbles among his fellow émigrés, a problem historically endemic in the Russian emigration. I recall trying to explain American democracy to a Soviet GI from the Ukraine, who could not imagine such a *bardak* (brothel or chaos) in his own country, a view that many Russians still share. While unhappiness with the Soviet system was a factor, the real catalyst for defection was usually something more mundane: a breakup with a wife or girlfriend, fear of disciplinary action for misconduct, or, in the case of Polish Central Committee member Seweryn Bialer, growing anti-Semitism in Polish leadership circles.

I am not sure that Uncle Sam got his money's worth from my interviews, but they certainly deepened my understanding of Soviet reality. A German scientist released from a Soviet concentration camp told me what seemed like an implausible story but one confirmed by Anton Dolgin, an American released from prison after Stalin's death: how the gulag camps were ruled by rival gangs, the *sukhi* and *blatnye*, deadly enemies, exploited by the prison administration to impose discipline. Dolgin described how he survived by being taken under the protection of the local mafia chief. And that notorious KGB informant in the diplomatic community in Moscow, Viktor Louis, told me how he had survived in Karaganda by arranging to become a hospital orderly.

**Life in Frankfurt**

Aside from the morning's chewable pollution from the Hoechst Dye Works, Frankfurt was a fine first post. We survived our first inspection and performed the minuet at the Steuben Schurz Ball. A German archeologist took us to visit the site of Roman barracks, which were ironically uncovered during construction of new American barracks. We tested wine on the Mosel and tried to understand the jokes at the months-long *Fasching* festivities, which begin on the

11th hour of the 11th day of the 11th month each year. By the time we left Frankfurt, our two children were bilingual in German, acquired at their Montessori school.

We preserved our peripatetic lifestyle. Christmas was spent at Lech in Austria, where we attended a magical midnight mass, with candles on all the graves, carols trumpeted across the valley from the church tower, and a sky filled with stars close enough to touch. One summer we joined our parents at Lorette de Mar, then a charming fishing village north of Barcelona. I returned to Spain for three weeks in 1956 to help prepare a questionnaire to be used with Loyalist Spaniards who were returning home after years in the Soviet Union. Their Russian wives were reportedly given a frigid reception by the Spanish ladies. I argued incessantly with former NKVD general Alexander Barmine, who was working for USIA, over the best approach to interrogation.

I was fortunate in having as a secretary in Frankfurt the wife of a consular officer Jim Carson. After Jim died in Haiti, Ginny joined the Foreign Service. Her first trial in New Delhi was to deal with an American streaker who ran through the consular section. For her children's first Christmas without their father, she met them at planeside on an elephant that proceeded to take them into town.

Back in Frankfurt, Consul General John Burns, an ardent Democrat, was so excited by the 1956 election that he woke us up at 3 a.m., urging us to come over and listen to the returns. We drank champagne and ate a kilo of black caviar provided by Chip Bohlen late into the morning. When we went to Paris with the Carsons to see John off after his tour was up, he invited us to dinner at the famous Tour d'Argent. Little did we know at the time that we would soon be living in Paris ourselves, though not eating at the Tour d'Argent.

## NATO in Paris

In August 1957, I was assigned as a Soviet expert to the Political Section of NATO in the Palais de Chaillot. The architect who had built what was supposed to be a temporary structure for the 1948 UN General Assembly explained, in answer to my ill-timed criticism, that the building was made out of pressed straw, a statement I can verify, having nearly fallen through the floor of the Archives.

This was a heady assignment for a junior officer. My English boss, Bill Newton, had started the BBC overseas service. Our major activity was the biannual review of trends in Soviet policy, conducted in French and English, inside a cloud of Gitanes cigarette smoke. To handle a particularly difficult delegate who was not a native English speaker, Bill deftly proposed language on one occasion that appeared to support his argument, when in fact it did the opposite. When Sir Evelyn Shuckborough, Anthony Eden's former secretary, became NATO political undersecretary, he was surprised to read a think piece that I had done on Soviet policy in the Middle East, saying, "I didn't know we could do this sort of thing." Since no one objected (perhaps because they paid no attention), I continued to write what interested me for distribution to NATO members.

With Sir Evelyn recently knighted, we nervously invited the Shuckboroughs to dinner. It was a disaster. Our chef had ordered a number of excellent wines from Nicholas and tried them all before dinner, so that he slipped and fell when serving the oysters and forgot to serve the guest of honor. Fortunately, Sir Evelyn, at least, was amused.

The years 1958–59 were tense years in Paris, with the Algerian militants of the Secret Army Organization (OAS) determined to topple De Gaulle for what they considered his treachery with his Delphic statement, *"Je vous comprends"* (I understand you), which masked his determination to abandon the fight for this *département* of France. Some of our French colleagues at NATO went into hiding because they were on an OAS death list. There were pairs of machine-gun toting police at each street corner in some areas. Together with Phil Valdes, our peripheral reporting officer at the Embassy concerned with the Soviet Union, I attended a 15,000-strong communist gathering, which marched out to the square outside the Val D'Hiver singing the "Internationale." Busy talking, we suddenly realized that we were alone with three rows of French CRS anti-riot troops descending upon us. Just flown over from Algeria, each looked as though he had eaten an Algerian nationalist guerrilla for breakfast. This was definitely not a moment for diplomatic posturing, and we fled.

The bursting pipes in our charming house in Bougivalle forced us to move into the city to an apartment on Quai Bleriot, across from the Statue of Liberty on the Pont Mirabeau, a fifteen-minute walk from the Palais de Chaillot. Our children became bilingual at their private school in Neuilly, Charles de Foucault. They also managed to fall into the Seine from a friend's houseboat anchored below us. Meanwhile, Nancy was exercising racehorses for the first time in her life near Chateau Lafitte. Sunday nights our Bistro Club composed of embassy officers and other Americans living in Paris, would try a new bistro among the few open on Sunday. The ritual was that the waiter would first bring a dish of tangerines for our administrative counselor Glen Wolfe's standard brown poodle, for his very French *mal de foie* (bad liver). We treated our own *foies* with a bottle of Sancerre.

I was surprised how rarely my French colleagues at NATO would invite French or other friends to their homes. If they entertained, it was in a restaurant. We were therefore quite touched when the director of our children's school gave us a farewell dinner. But it was another disaster. Nan was allergic to oysters but too shy to refuse a first course of *colliquages*. Violently sick, she spent the evening upstairs in bed.

We had been amused when the school director had expressed amazement that we were Protestants. "Why, you are so gay." We came to understand the remark when we tried to help good French friends at the wedding of their daughter at Aigre near Bordeaux. Their daughter had become a typical American bobby-soxer during a year at UCLA, where she met a man from South Dakota from a fundamentalist Methodist family that did not drink or appreciate good food.

"There are civilized Americans," her parents commented. "Why did she have to fall in love with this one?" I served as interpreter at the ceremony by the Mayor of Aigre. Then it was agreed that there would be a Protestant service at the Huguenot *temple*. Catholic friends swarmed in, delighted for an excuse to visit this curious institution. The pastor, in turn, was delighted to have all these heretics in his temple and delivered an endless fire-and-brimstone sermon, with no relation to the marriage. Happily, the uncle of the bride owned a cognac distillery where we relieved the tension by testing cognacs dating from 1860 to the 1930s. (I think we decided that 1904 was the best).

Many years later I made the mistake of recalling to a local jeweler in Isle sur la Sorgue near our house in France the history of nearby Coustelet, where the Huguenot population was massacred by the Catholics when they opened the gates of their besieged town with the promise of being spared. The rage of the jeweler, an ardent Huguenot, over this event many centuries before, illustrated the continuing sense of victimization among this Huguenot minority and the lingering problem in Europe of too great a historic memory.

## Battle for the Third World

The embassy and I had assumed that I would replace Valdes as peripheral reporting officer, but Washington had other ideas. Our "cold warriors" and Congress had panicked when Soviet General Secretary Nikita Khrushchev decided to challenge America globally by reaching out to the developing countries, offering them a vision of progress and aid. Neither the Russians nor we fully appreciated the determination and skill of these recently decolonized countries at playing the great powers off against each other to protect their newly achieved sovereignty and extract more foreign aid. Visualizing a red tide sweeping across the world of starving millions, policymakers in State decided to establish a "short- haired" policy planning group drawing together specialists from the major geographic areas to devise strategies for countering this communist threat.

I was the Soviet expert assigned to this new unit, U/CEA (Communist Economic Affairs). My colleague Middle East expert Carleton Coon received a typical bureaucratic, cold-shoulder response when he tried to work with the Near East Bureau: "We will handle our own Communists, thank you." But Philip Habib and I, assigned to Africa about which neither knew anything, were able to take advantage of the explosion in newly independent nations that had quite overwhelmed the Africa Bureau.

I can still recall visiting the Congo desk officer, Tom Cassily, to discuss messages warning that Soviet officials were making overtures to the Congolese nationalist Patrice Lumumba and the Belgian communists. In his trademark rumpled blue pin-stripe suit he was exhausted. I can still hear his acid comment: "I read the same damn cables that you do, and what do you want me to do about it?"

Well, we did do something about it, writing a policy recommendation that the United States try to use the United Nations as a firewall in the Congo to avert an East-West confrontation. I only later learned that Ambassador Clare Timberlake in the Congo had made a similar proposal. As senior officer, Habib presented our proposal to Secretary of State Dean Rusk. According to Habib, the secretary asked what this would all cost, and he gave the secretary an arbitrary figure of around $200 million. As a streetwise Lebanese Brooklynite, Habib knew that all the secretary needed was a figure to present to Congress.

What we judged to have been "our policy" served its purpose, and confrontation was avoided, but the Congolese who came to power did little to modernize their new country. To be fair, the Belgian administration fled the Congo in panic, leaving only some twenty-six Congolese with partial university training to govern and modernize a country two-thirds the size of the United States The U.S. Congress became increasingly critical of President Sese Seko Mobutu and questioned the value of having the CIA station chief in Kinshasa, Larry Devlin, doling out money regularly to keep "our bastard" in power. It was easy to criticize, but finding someone with the personal authority and skill to hold the Congo together, with its bitter ethnic rivalries, was not self-evident. Under the successors of Mobutu, like the most recent, Joseph Kabila, the Congo descended into civil war.

I know that USAID officers tried hard to make a difference in Africa, but their aid was not always well conceived. In the Congo, the Belgians used to maintain the roads by requiring that each village take responsibility for its own *tranche* of road, using hand labor. But USAID set up a huge depot for road-building equipment that proved a waste of money, because the Congolese did not have fuel to power the Caterpillars nor the skill to service the equipment.

### African Seminar

When I heard about a seven-week seminar touring Africa, I grabbed the opportunity to see the Black Continent for myself. At our first stop in Dakar, Senegal, where students boycotted us because of racial incidents in the American South, I was awakened the first night by a pencil light near my bed that turned into a thief running out with my pants over his arm. He had already neatly stacked the wallets and clothes of my CIA and DIA roommates on the terrace. In Kaolack in southern Senegal, we were impressed by the local governor, who spoke unaccented French, and a Red Cross ball marked by beautifully dressed Senegalese women, chaperoned by their formidable mothers.

In the Gambia we were invited for dinner by the governor, who had just recovered from an accident where a Portuguese Man of War jellyfish came between him and his surfboard. Our stiff colonial host in evening dress did not encourage conversation. We wondered why this former governor of Kenya should now find himself in the Gambia. Later one English officer said: "If you Yanks would like the Gambia, you can have it cheap."

From the Gambia we went to Guinea, where we were invited to see the Ballets Africaines. With the labor officer's wife on my arm, I was hesitant to push my way through a mob of drunken Guineans armed with Kalashnikovs until a huge Guinean grabbed me and pulled us into a vast barnlike hall. Seeing Caucasians looking for a seat, the radical President of Guinea Sekou Touré had two members of his cabinet give up their seats. The rainy season started when we were in Conakry, Guinea's capital, and in minutes there was a foot of water in the church where we were attending a service, and everything smelled of must.

From Conakry we flew to Ouagadougou, capital of Upper Volta (now called Burkina Faso). Our guide, a young Mossi from the local ethnic group attached to the French-founded IFAN Institute of Sociological Research, took us to the village where he had been born. He said that the wheel had not come to his village, which as a result lacked all the trash of more modern villages and still had a little old lady brushing a clean dirt floor with a twig broom. In effect, this graduate of the Musée de l'Homme in Paris had bridged a gap of 2000 years in his lifetime. I will always remember the sight as we drove out of Ouagadougou—a single file of lovely, slender women with jars on their heads, silhouetted against the rising sun. We were shocked to hear before we had even finished our tour, that the strapping former Marine in charge of our embassy had died of

either typhus or hepatitis—hardly surprising with the cattle graz-
ing in the pond that supplied the city water.

When we complained of the noise at the hotel where we stopped
on our way to Accra, we were moved to a hotel with perfumed
sheets that appeared to be a local brothel. These were the days
when Kwame Nkrumah, the graduate of Lincoln University who
was president of Ghana, sought aid from both the United States
and the Soviet Union in his drive for pan-African union. We flew
on to Lagos, Nigeria, in a Soviet plane with Soviet crew. Aside from
the Benin bronzes in the local museum, what impressed me most
was the heavily accented English spoken by Nigerian officials, in
contrast to the fluent French of the Senegalese.

En route to our major destination, Makerere College in Kampala, Uganda, we stopped in the Central African Republic, known
for its butterflies and diamonds. In Kampala, it had been arranged
for us to spend the evening with local families. I was lucky to get a
Pakistani barrister with two beautiful lawyer friends, obviously vy-
ing for his attention. He had just lost an election when he had tried
to persuade his fellow Pakistanis in Uganda that "we are Africans
and need to integrate into Africa." On a visit to a tea plantation, I
asked the Pakistani manager what was being done to train Africans for managerial positions, and he replied, in effect, "They are
not worth the effort." The Pakistanis paid for their arrogance by
being expelled from East Africa shortly following our trip. In the
Ugandan Parliament, which replicated the Parliament in London,
we tried to understand the members' wearing their white wigs and
roaring with laughter during the traditional question period.

At Makerere we were lectured to by a Scot, who had just returned from visiting the jailed Jomo Kenyatta, reputed head of the
feared Mau Mau fighters against the British. The Scot had been tre-
mendously impressed by Kenyatta, who later made him, as I recall,
his minister of agriculture. The White Highlands, with their coffee
plantations, golf courses, and charming inns with barefoot waiters
serving steak and kidney pie were a very colonial experience.

On the way home, via south Yemen, I was amused to see a
mother in Aden pointing out to her child the lawn outside our Consulate, the only green spot in this town of rock. In Greece, having
seen *Never on Sunday* with Melina Mercouri, I tried to find Piraeus.

Where the streetcar left me off, there was no sign of a port, and I was starving. I was getting nowhere explaining my predicament to the owner of a small, deserted bar until, in desperation, I spoke Russian and found that the bartender was a homesick immigrant Greek from Soviet Odessa.

Back in Washington, I was promoted as a Class-3 officer to the position of supervising intelligence research specialist working in what was now called INR—a bureau—rather than OIR, the Office of Intelligence and Research. There I set up interagency meetings of specialists to look at real or potential crisis areas and suggest what the United States could or should do. For me, this was a most interesting operation, but, frankly, it acquired little traction.

# PHOTO GALLERY 1

Young Tom with stepfather Prince
Seesodia and dog Pat, Lake Simco,
Canada, 1932

Tom's mother, Katharine Winterbotham, and Prince Seesodia
on the Riviera in 1930s

Tom (standing at left) on Stanmore Park (England)
Cricket team, 1937

38

Navy Ensign Thompson Buchanan, 1945

Tom and his bride Nancy
Shaw at their wedding, July
1945

Prince Seesodia toasts Nancy at the wedding

Tom's Mother and sister-in-law Charlotte Burchell at the wedding

Tom on top of Grand Teton, Jackson Hole, Wyoming, 1948

Nancy gives daughter Barbara Buchanan her first riding lesson, circa 1950

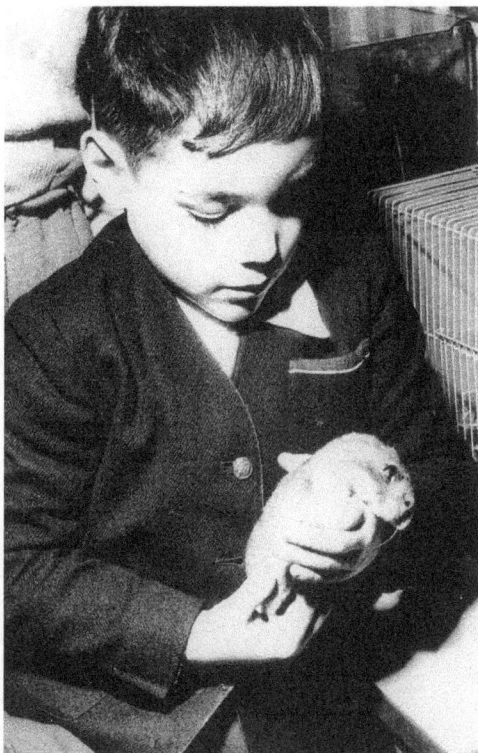

Campbell Buchanan and his hamster, Frankfurt am Main, Germany, 1956

Nancy (seated center) with the minuet team at Steuben Schurz Ball, Frankfurt am Main, 1956

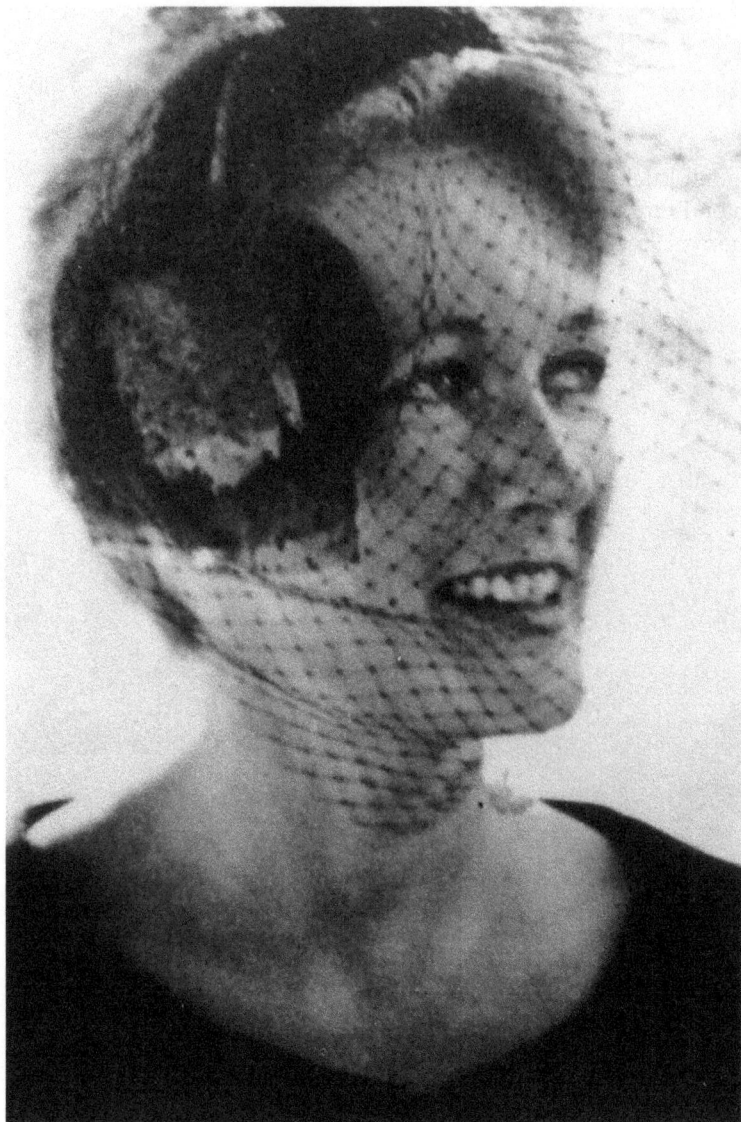

Nancy in Paris, 1959

# 6

# Embassy Moscow: a Time of Crisis

In September of 1962, I was assigned to my first regular Foreign Service post at the Moscow Embassy, replacing Adolph "Spike" Dubs as head of the Foreign Political Section. (Spike, a wonderful human being, was later assassinated when he was ambassador in Kabul). It was a fascinating though difficult time to be in Moscow. Though the Cuban Missile Crisis threatened the survival of our two countries, we were probably less tense than Americans back home because we were sheltered from the alarmist reporting of the American press. And my past work had convinced me that the Soviets would be the first to blink

Ambassador Foy Kohler limited knowledge of the crucial exchanges of messages between Kennedy and Khrushchev to his deputy, his political counselor, and some translators. Being excluded from this close inner circle was probably a good lesson for a somewhat cocky, relatively junior first-tour officer.

The next crisis was the assassination of President Kennedy. We were having dinner at the French commercial counselor's when an Agence France Presse correspondent was called out to take a message. Shocked by the news and queasy from trying to smoke a large Cuban cigar with sophistication, I rushed to the Embassy. Not only did Khrushchev himself come to sign the condolence book at the Embassy, but Russians on the street would stop us and say, "How could you have allowed this to happen?" For to many Russians, Kennedy was the symbol of the young leader they would like to have had, and they assumed he was destroyed by "reactionary circles opposed to peace."

I opened my tour with a crisis of a different dimension. Accompanying Bill Morgan, our book procurement officer, on a trip to Baku, Yerevan, and Tbilisi to buy books under our bilateral agreement with the Russians, we found the atmosphere increasingly tense. In Baku, which had its first snowfall in twenty-five years, our "tail" slipped and fell on his face while following us up the long staircase above the city. On the flight to Yerevan, the stewardess apologetically hung coats all around us so that we could not see out. We were, however, allowed to travel from Yerevan to Tbilisi by a train that passed for miles along the plowed stripes and watchtowers along the Armenian-Turkish border. The director of the Folk Theater in Tbilisi, whom we met on the train, failed to meet us as he had promised. We understood why when we read one of the wall newspapers on the hill above Tbilisi. It showed pictures of U.S. embassy officers expelled as CIA spies for running Colonel Oleg Penkovsky, the highly placed Soviet officer who had explained to us how the Soviet missile program was much weaker than we believed. It was this information that had made it easier for us to call Khrushchev's bluff during the missile crisis.

**The Khrushchev Era**

Back in 1952, while in the Intelligence Bureau, I was scheduled to be assigned to Moscow, but the money ran out. Sovietologists who had served in the Soviet Union when Uncle Joe was still alive considered themselves *la crème de la crème*. Working in Moscow when Khrushchev was secretary general was the second best thing, for it was a period of relative détente, and Khrushchev was such an ebullient, sociable, intellectually curious and unpredictable personality, with a gift for pithy, unbureaucratic language. His much-quoted "We shall bury you" reflected his confidence that the Soviet Union would defeat the United States in the competition between two rival visions of society.

With his impulsiveness and lack of good judgment, Khrushchev was his own worst enemy. In trying to emulate the success of American agriculture, he decreed that corn should be planted across the country, regardless of the suitability of the region, earning him the nickname of *Kukuruznik,* the Corn Man. His campaign to convert the virgin lands of Kazakhstan to wheat production was a disappointment. His efforts to restructure the party bureaucracy and to force Soviet peasants off the land into urban *Agrogorods* alienated the party bureaucracy.

But his most serious mistakes were in foreign policy. His "liberal" policies were blamed for the uprising in Hungary in 1956 and the

deterioration in relations with Communist China. His most griev-
ous mistake was his gamble that he could strengthen his bargaining
position on issues like Berlin and deter an American attack on Cuba
by emplacing nuclear weapons on Cuba. The result must have infuriated the conservative military, humiliated by being forced to back
down. When I left Moscow in August 1964, Khrushchev had had
a good harvest, in contrast to recent times, and analysts thought
that this had strengthened his position. But in October a coterie of
party officials led by the future general secretary, Leonid Brezhnev,
and KGB Chairman Vladimir Semichastniy took advantage of his
absence to depose him and return "order" to Soviet society.

**Russian Humor**

A lot of humor went out of Russian life with the departure of
Khrushchev. A feature of that period of thaw was the assortment of
jokes that came to be known as Radio Yerevan. For April 1 the embassy used to put out its compilation of the best Soviet jokes, most
of which, unfortunately, I cannot remember. One of the jokes concerned that well-known Armenian proclivity for political survival.
Politburo member Anastas Mikoyan was the prototype of a surviv-
ing Armenian. In one of these jokes, the tsar is back on the throne of
Russia and Khrushchev is an old man living in exile. Khrushchev
calls up the tsar and pleads to be allowed to return and to die in his
*rodina* (motherland) like any good Russian. The tsar hesitates and
then says, "Wait a minute, I must consult Count Mikoyan."

On the morbid side was one of the jokes at the height of atomic
warfare fears: "What do you do when the air raid sounds? Answer:
Find a white sheet and walk, not run, to the nearest cemetery."

Sometimes the humor was unintended. On my first taxi ride in
Moscow in 1962 I asked the driver about life today and under Stalin. "Oh it was much better under Stalin," he said. "Vodka cost only
two kopecks a bottle."

Happily the Russians have not lost their gift for sardonic humor. On my last trip in 2010, our guide said that Russians see a
virtue in crossing the road against a red light; "It gets rid of old
fogies." The Russian economy, she said, is referred to not as the
"free market but the flea market." With their usual optimism, when

asked, "How are things?" they answer, "Better than they will be tomorrow." A pessimist, she said, was an "informed optimist."

In one of the many jokes comparing national character, the Russian is said to be" the person who is naked but has one apple and thinks he's in Paradise." In Yaroslavl, our guide kept pointing out unfinished local projects that, he said with a smile, will "of course be ready for the millennium celebration in September. The city government has said so."

## An "Asia Expert"

The embassy did not have area experts in 1962 as we did later, so I took what seemed the most interesting region, Asia, as my sphere of responsibility. Much of my work was initially focused on Laos, where efforts to set up a neutral, independent coalition government under Prince Souvanna Phouma had broken down, with Souvanna Phouma siding with the Communist Pathet Lao against the Royal Lao government supported by the United States, Thailand, and the CIA-trained Hmong mountain tribes. In our NATO counselor meetings in Moscow, I would discuss the Soviet position on Laos, based again on patterns of Soviet behavior and analysis of the press. At one point Ambassador Kohler said, "I hope to Hell you know what you are talking about." Happily events proved me right. When Governor Harriman came to Moscow to try to persuade the Soviets to withdraw their support of the Pathet Lao and Chinese, I was amused, as the note taker, to see the governor ostentatiously remove his hearing aid after determining that Foreign Minister Gromyko had nothing new to say. At the luncheon he gave Harriman in the Foreign Ministry guesthouse, Gromyko was highly amused when one of Harriman's aides told him that the governor's nickname was the Crocodile. The Crocodile was not amused.

As the self-appointed Asia expert, I enjoyed the company of the sophisticated Laotian ambassador Kamphan Panya, who was later forced to flee Laos. The details given by the Burmese chargé about the sex life of *Homo Sovieticus, as* seen, he claimed, through the eyes of Burmese students, became the basis of a dispatch that was never approved for transmittal to Washington.

It was also the Burmese chargé who invited us to dinner at the appropriately named Winter Garden at the Prague Restaurant, which had no heating when the temperature outside was in the low teens. The poor Asian ladies shivered in their beautiful Saris. The wife of the British minister held a bowl of hot soup to her bosom, and the Soviets resolved the problem à *la Russe* by drinking more vodka.

## Life in Moscow

Living at No. 45 Leninsky Prospekt among Russians and not in the Embassy Ghetto gave us a better feel for Moscow. Walking our first night in the huge adjacent Gorky Park, we had a charming introduction to Moscow from a scruffy-looking Russian who ran over, bowed deeply, and congratulated us on the coiffure of our little black poodle Girouette, which had been recently coiffed by her former owner, the Princesse de Nouilles, in Paris.

In wintertime the paths in Gorky Park were turned into a great ice-skating rink, with lights and music. In the summer, our visiting daughter and a French friend were snuck into the restaurants in the park by their Russian boyfriends. Closer to the Kremlin, what was perhaps the world's largest circular swimming pool had been built on the site where in the 1930s Stalin had blown up the largest cathedral in Moscow, the Church of the Redeemer. As one of the Russian-speaking fathers, I found myself one day with the two small boys of the deputy chief of mission, one under each arm, as they received a swimming lesson. With the temperature outside in the low teens, the water from the heated pool congealed on our upper bodies, turning us into snowmen. (In 2006, Nan and I attended a service in the now-restored cathedral, with the church hierarchy resplendent in their rich robes.)

In summer, we would spend weekends at the embassy *dacha*, or country house, in Tarasovka, about ninety minutes outside of Moscow. Or we would go to the beach in a large reservoir near Moscow to watch the crowds of young people swimming, boating, listening to the songs of the dissident artist Vysotsky, or taunting the local guardians of law and order by dancing the forbidden Twist in the sand.

We were spoiled by the opportunity to hear the Borodin Quartet and operas like *Boris Godunov* and *Eugene Onegin* and to see the great ballet dancers of the day perform to knowledgeable and loving audiences for a ridiculously low price. The tour of the Robert Shaw Chorale singing the Bach B Minor Mass, Monteverdi Baroque music, and American spirituals was a "happening." Only music majors had ever been allowed to hear the Mass, and one painter was so moved that he painted a picture of Christ and gave it to Shaw at the next day's performance. While the female voices in those days tended to be strong but shrill (in contrast to the wonderful singers who have emerged in the new Russia), Shaw was impressed by the male basses. He declined, however, the invitation to become a director of the historic Kapella choir in Leningrad.

## Touring European Russia

From Moscow, we would travel as much as travel controls allowed to the different historic sites: the centers of Russian princely power before Moscow, in Vladimir, Suzdal, and Rostov Velikhi.

Our most memorable trip was to Yaroslavl over a weekend with three friends. After leaving the historic site of the Russian Orthodox Patriarchate in Zagorsk, the limitless white landscape suddenly dropped off without any warning, and I found myself heading down a steep, icy hill with a huge hole in the middle. To avoid breaking my axle, I braked and skidded around. Then my Checker station wagon slid slowly off the road upside down, fortunately into a snow bank far below. The top of the Checker was flattened but only one lady passenger suffered some broken ribs.

Eventually our faithful KGB came to see why we had not arrived at our next checkpoint. A polite "private citizen" promptly stopped a mail truck so that we could load our baggage and passengers for a return to Moscow, while I stood on the road, shivering from shock and from the minus 26 degree temperature, while the police pondered for an hour how the accident happened and how to get out my car. Finally I was taken to the police station at Pereslavl-Zalessky nearby, famous for having the pond on which Peter the Great sailed his first boat.

Hearing that there was an American in town, the mayor hurried over to discuss in depth his favorite writer, Mark Twain. Boy, was he disappointed! All I could faintly remember was Huckleberry Finn. I froze in the back of another car of "friends" returning me to Moscow. I sent for my old Ford station wagon, which had not yet been sold in the States, and sold what was basically a solid Checker motor and chassis to a collective farm, which is probably still driving the monster.

My next trip was memorable for a different reason. I was traveling alone by train to Budapest to join the family on vacation, convinced that the Soviets would try somehow to compromise me. In fact, in Kiev I watched as a gorgeous blond walked down the platform and headed for my compartment. She claimed that she was traveling to join her military husband in Budapest. I waited for her to make some approach, but she was very proper, asking me to leave as she prepared for bed. I concluded that the lady was probably legitimate, but had been asked to report back on my reaction to the situation.

The Soviets were not enthusiastic about our efforts to measure conditions in the hinterland against those in Moscow and Leningrad. For protection against provocation, American officers had to be accompanied by at least one fellow officer, family member, or diplomat from another NATO mission. Even so, we had incidents, usually involving the military. We were briefed on what to look for on our travels, such as the cost of food in the markets, or the license plates of military vehicles to help our military identify Soviet military units, which seems so silly today.

## Touring Central Asia

Our most exciting trips were certainly those to the capitals or historic sites of Central Asia. We were among the first diplomats to be allowed to visit Khiva in Uzbekistan. This walled city, with its adobe buildings, was what I imagined a city in biblical times would have looked like. Standing on the walls, we could look down into the garden of a two-story house marked by a beautifully carved wooden pillar joining the patio to the first floor. Even though there was no hot water, and the facilities were out of doors, we always cited our hotel in Khiva to surly Russian hotelkeepers as a model of hospitality.

In Tashkent, we experienced the nationalist pride of a people with an ancient history. At the beautiful National Museum, the curator, looking for emphasis at our accompanying Russian guide, pointed out how old the Uzbek civilization was. Older, he implied, than Mother Russia.

In Samarkand, the oldest city in Central Asia dating back to the third century B.C., we saw workmen trying to replicate the extraordinary blue of the tiles on the great mosques. The Uzbeks pointed with pride to the celestial observatory of their early astronomer, Ulan Beg. In Bukhara, we admired the superb brickwork and chased down a famous former mosque, which seemed to have vanished until a small boy showed us the ruins. We also visited a synagogue of Sephardic Jews who had not yet emigrated.

We timed our visit to Frunze, now Bishkek, capital of Kirghizstan, for our son's Easter vacation, but it was ill timed. Apparently some Communist conference was in progress, and security was very tight. When I visited the bookstores I found them all closed for "inventory." We were delighted to be invited by the Imam, the Muslim religious leader of Kirghizia, for some "real food," meaning not Russian. But when we arrived for dinner after a day of harassment, we found four "goons" waiting in a car outside the mosque. The Imam was clearly relieved when I suggested that we postpone our dinner, and he gave us the name of the best restaurant in town, which had a great band. While we were eating, accompanied by our son's red-haired teacher from Moscow, a drunken Kirghiz officer staggered past our table and hit one of the forty Iraqi Air Force trainees seated behind us. To a man, these bourgeois-dressed Iraqis exploded from their seats with knives in their hands, and it was a miracle that the Kirghiz got out alive. With all of his charges chasing the Kirghiz, the Soviet group leader turned his attention to our ebullient redhead. We told him that we were going to the Easter service in the local cathedral, and he offered to join us, claiming that he was a former Interior Ministry (NKVD) police officer who was studying to become a surgeon in Frunze.

We arrived at the church in the pouring rain in mud a foot deep. With nothing to do on a Saturday night, all the young hooligans in Frunze were trying to push their way into the church. Sturdy Russian *babushky* were pushing them back down the long flight of stairs, and our dubious new acquaintance and I came to their support. Suddenly the Russian pulled our redhead with him into the church, where she had the impression he was truly interested in the service. Left alone, all the women turned on me, assuming that I was a police officer, saying, "Aren't you ashamed of yourself allowing this hooliganism?" When I explained who we were, they crowded around, shielding us with their umbrellas, asking if it was true that in America we broadcast the Easter service on television—an amazing question from someone near the isolated Afghan border.

Interested in a portion of the Kirghiz border claimed by China, I grabbed a taxi on the street and asked the driver to take me to beautiful Lake Issik-Kul. If I had known that the Soviets used the lake for secret experiments with torpedoes, I might have been more discreet. As it was, we did not get far. A high KGB officer stopped us on the outskirts of Frunze, took our chauffeur's driving license, and told me I could not go to the lake. When I tried to bluff, pointing out that the area was not clear on the Soviet map of travel restrictions, he said crisply, "Take it up in Moscow."

Finally we were given permission to visit a factory producing heavy, felt white and black Kirghiz hats, but when we got there the director denied having received any instructions. So all we could do was watch as the workers leaving the factory were frisked to see that they were not stealing any material. Back in the hotel I had the satisfaction of complaining loudly about the Soviet pretensions of wishing to encourage tourism and hearing my complaint echoed by some bystanders.

I was amused that, when I had my hair cut, my lady barber called all the other barbers over to feel my hair, "just like that of a girl," so different apparently from the coarse hair of the Central Asians.

Our trip to Alma Ata, at the time the capital of Kazakhstan (which is now Alana), stands out in my memory for different reasons. It was supposedly the apple center of Russia, but I do not recall the apples. We were intrigued instead by a brochure advertising skiing, with a pretty girl in a bikini carrying skis. And so we began hiking to where we were told there was skiing. We only turned back when we found we were in a military restricted zone with the ski slope visible but clearly lacking in any ski lift. A typical example of Soviet out-of-phase planning: it is easier to issue propaganda than build a ski resort.

On the way down, hiking alongside us was a man with a striking resemblance to Leon Trotsky. On the bus he picked up conversation with us in rusty but idiomatic English that he said he had learned at the London School of Economics in the early 1930s. He was, he said, the overseer of the Baptist churches of Central Asia. He could have been a tough communist serving a different cause. He lost interest in us promptly when we told him that unfortunately we could not attend his church service that evening because we were going to the theater.

A second night we went first to a local football game that almost turned into a riot among the spectators, and then to a restaurant where we found ourselves stranded by a torrential downpour and no way to get a taxi. Finally a police paddy wagon drove up and invited us to get in. We laughed uneasily as the burly cop in the front seat, turned and said to us through the bars of our cell with a grin: "You won't get out of here in a hurry." It was a shock at our hotel when its honored foreign guests were delivered in a paddy wagon.

# 7
# Africa Calls: Destination Burundi

In what we called the "April Fool's" list, each of us was asked to name our choices for the next assignment. When I asked for a Francophone sub-Saharan post on the water, it was like asking to move to Harlem when one was living on Park Avenue (read: the European Bureau). My friends thought that I had lost it. But my fascination with this new frontier took precedence over bureaucratic calculation. Personnel gave me my choice. Though not exactly what I had in mind, Bujumbura, Burundi, was French-speaking, in one of the most beautiful countries in the world, and located next to the Congo on Lake Tanganyika, the longest lake in the world.

After serving on a promotion panel in Washington, agonizing over officers with efficiency reports inflated by their well-meaning superiors, I arrived in Bujumbura in December 1964. I quickly realized that this was a very small town. When I picked up a student on my way to the university, he said promptly: "Why, you must be Mr. Stanger's (the DCM) replacement." And when I went to the Paguidas Bar that night, where everyone was reading the Communist Chinese Information Bulletin and I was the only Caucasian, a man came up to me and asked how I liked Burundi, claiming that we had met in Moscow.

One should not speak ill of the dead, but Ambassador Don Dumont and his *pied-noir* French wife from Algeria were very difficult. In our eleven-man post, under crisis conditions, I remained Mr. Buchanan. No one was invited aboard Dumont's boat, and he insisted that he be the only one to have contact with high Burundi officials, which complicated life for me when I was chargé. Perhaps the Dumonts felt threatened by an attractive, fluent-French-speaking couple intruding into their African domain from the European

Bureau. I might question Don's judgment but he was an excellent writer, synthesizing the complex politics of Burundi as *un panier de crabbes* (a basket of crabs).

The first fortnight the Buchanans were in heaven, swimming in the surf of Lake Tanganyika between the high Congolese mountains on one side and the flower-covered hills above Bujumbura on the other. We tried to picnic in the countryside, but gaunt Tutsis standing by us destroyed our appetite; and we got into the habit of taking the embassy's unseaworthy motorboat to a small spit of isolated land. That is, until a Belgian resident told us that the little stream nearby was the best area for bilharzia (or schistomiasis), a snail-carried parasite that attacks the kidneys, and that it was also a favorite area for crocodiles.

Meanwhile, the cannonading from the Congo reminded us that the civil war there was in full swing. And we remembered the rumor of a coup we'd heard when we left Washington. Our world dissolved in our third week when the moderate Hutu prime minister Ngendamdumwe was assassinated. The moderate Tutsi king Mwami Mwambutsa I then expelled the Communist Chinese, who used to hand out their weekly subsidy to radical Tutsis every Thursday.

This curious alliance of Communist Chinese and the Tutsi, in effect a feudal aristocracy that had ruled over the Hutus for centuries, has an easy explanation. In 1959, the Hutus rose in Rwanda and massacred the Tutsis, literally cutting this statuesque race down to size. With thousands of Rwandan Tutsis in refugee camps in Burundi, the Burundi Tutsi elite were understandably paranoid. The Tutsis blamed the Belgians as the Mandate Power for Rwanda-Urundi and, by extension, their NATO allies, for the 1959 massacre and considered the "enemy of my enemy" to be their friend.

The Communists were quick to exploit Tutsi radicalism. Cuba's Che Guevera visited the refugee camps to recruit Tutsis for the Congolese civil war, and insurgency-training films for Congo-bound recruits were reportedly shown in the Paguidas Hotel.

Our embassy was directly affected by the assassination because the authorities arrested our leading Tutsi local employee and tortured him until convinced that he was not directly involved. It emerged, however, that he had been a member of a radical Tutsi organization and also had had his hand in the embassy till. Then we entered a period of martial law leading up to the parliamentary elections in the spring.

This did not prevent us from going about our business. With the ambassador's $25,000 Special Fund, I managed to buy a truck for a farmer's market and was working on a rice project when our world fell apart.

## Missionaries in Burundi

Even though we had missionaries on both sides of our family, we had not appreciated how active they still were in trying to help people with their physical as well as spiritual needs in Africa. We were particularly impressed by the Catholic effort to reach out to Africans, including adapting their hymns to African rhythm.

With a wonderful Seventh Day Adventist missionary called Red, we attended the consecration of the first Hutu bishop in Kitega. Our friend then took us to Rwanda, where we witnessed a hysterectomy operation performed by a Canadian Seventh Day Adventist doctor in a tiny clinic, where he first had to get the generator going for the lights. The patient had decided after three years that the lump in her stomach was not a baby but a huge tumor. The normal supplies for such an operation were lacking but the patient survived. This very unmissionary-appearing, nattily dressed Canadian would go back each year to Canada to raise money for his clinic.

On this same trip, we drove in our Jeep from Kagera National Game Park along a ridge in Northern Rwanda, which led to some forty volcanoes towering over small lakes on one side and Lake Kivu on the other—breathtakingly beautiful. Indeed, if ever Rwanda and Burundi get their act together, they should become among the greatest tourist attractions of Africa.

## Pyrrhic Hutu Victory

Back in Burundi, the Hutu had won a resounding victory at the polls. I was not privy to discussions between our CIA station chief and the ambassador, but I strongly suspect that we had a hand in funding the more moderate Hutu politicians who won the election.

Unfortunately, some Hutus, like the Hutu ambassador to the United Nations, reacted arrogantly to the Hutu victory in a way that only confirmed the worst Tutsi fears. The Tutsis responded by moving aggressively against anyone suspected of being pro-Hutu.

Some missionaries could be very stubborn. In the case of one Protestant missionary family with twelve children, I tried and failed to persuade the father not to go back into the Congo, which was in the throes of civil war. Particularly since he lived in the area of Stanleyville, which was wildly anti-American following the landing of American paratroops to rescue members of the U.S. Mission.

In their paranoia, the Tutsi radicals read something sinister into every action of our mission. On one occasion when I was chargé, one of the frequent erratic curfews was announced as guests were due to arrive for a farewell party at my residence for a helpful administrative "rover." It was too late to call off the party, so I decided to go ahead. After a bit, my residence was surrounded by drunken soldiers with Kalashnikovs, who would not let anyone leave. It finally took the intervention of Minister of Defense Micombero, Burundi's future president, after he had drunk all my Scotch, to flush out the mayor, who lived across the street and who gave us permission to leave at 5 a.m., when the curfew was due to end anyway.

When I tried to drive one of our terrified Tutsi local employees back to his house in the African quarter, we were stopped by soldiers, who were infuriated by something our local said to them and took us to the police station. There a huge, bull-necked Hutu said to my trembling Tutsi, with a leer: "What makes you think I will hurt you?" When the mayor denied having given me permission to be on the road, I told the police chief that I would be back the following morning and expected to retrieve my local unharmed. To avoid the very nasty experience of spending the night in a Burundi jail, I would ignore calls by drunken soldiers to halt, driving slowly with hairs raised on the back our necks, praying no one would shoot. It later appeared that the mayor thought my party was to celebrate the Hutu election victory.

## Tutsi Radicals Seize Power

Exploiting ill-advised statements by overconfident Hutus, the radical Tutsi ministers were reported to have gone to the king and other more moderate Tutsis arguing that they would have to act together to save themselves from Hutu ethnic cleansing. It was our impression, at least, that the Tutsis then behaved with their traditional guile and encouraged the Hutu-led police to attempt a coup. The fact that the army arrived almost simultaneously with the Hutu police at the palace suggested that they were set up. The police then retreated to the army camp behind our house. We were awakened at 2:30 a.m. by the sound of gunfire around us and lay down in the corridor outside our bedroom. Having created the pretext, the radical Tutsis then rounded up every Hutu of any prominence, driving them three nights in a row past our residence to be machine-gunned in the central stadium, where I was told that the Hutus, often from mission schools, died singing Catholic hymns, with their bodies left there *pour encourager les autres.*

Civil war spread, with fighting reported between the Tutsi military and the Hutu population across the country. There was panic in Bujumbura as rumors spread that an army of Hutus, supported by the feared Twa pygmy warriors and their blowpipes, historically allied with the Tutsis, was advancing on Bujumbura. The Belgian military began plans to protect and evacuate their nationals.

Our greatest concern was the safety of our missionaries in the interior. As noted, the Catholic missionaries were linked with one another by shortwave radio, but not our Protestants. I would spend hours in the central radio station trying to contact those American missionaries who had radios. I was particularly concerned for an attractive Episcopalian couple and their two daughters, miles in the interior. When they finally emerged, taken by the army to Bujumbura, they had a horrifying tale to tell.

Tutsi troops burst into their house one morning, terrorized the children, and searched the house, finding one rifle and a hunting bow. The minister had tried to register the gun, given him by a friend, but was given a runaround. More serious was his hunting bow. Clearly, the Tutsis told him, you have these weapons to help arm the Hutus. No white man uses a bow and arrow. When he insisted that he had bought the bow to hunt game in Burundi, they pointed at a small door in a distant house and told him to hit the door. He had actually never used his bow, but he prayed, he said, and drove an arrow through the door, half-persuading them that he was telling the truth. The family was taken under armed escort to Bujumbura, where they were, in effect, placed under house arrest, pending trial. Rather than risk the outcome in the hysterical atmosphere of the time, we arranged to fly the family out of the country.

The missionaries could be stubbornly unreasonable. I remember trying to persuade one Protestant missionary with some twelve children not to return to his mission in the Congo in an area where civil war raged and the radicalized Africans were angered by the landing of American paratroops in Stanleyville to rescue captive members of the U.S. mission. He adamantly insisted on returning with his family. I never learned what happened to him.

In the radical atmosphere that followed this coup, there were threats to kill the ambassador, our USIA director, or me. We did not take the threats too seriously. Still, I told the ambassador that I did not think it particularly wise for him to ride to work on a bicycle, but he insisted he was not going to be intimidated. The Tutsis spread rumors now that our daughter, who had worked digging latrines for Tutsi refugees in the hills, was not really my daughter but a spy. In preparation for entering Sarah Lawrence, and before the radical coup, daughter Barbara had insisted on having the experience of being one of a few Caucasians in the University of Bujumbura. She was protected from serious hazing by a muscular Hutu student whom we had befriended. Her hazing involved writing an essay explaining why Burundi should reestablish relations with Communist China and having to crawl across a local street singing the Communist *Internationale*. We fear her Hutu friend did not survive the subsequent massacres.

Once when I became chargé, I tried to stem this deterioration in relations by going to see the minister of education, reportedly the most radical and intellectual of the Tutsis. He advised my ambassador to speak with Minister of Defense (later President) Micombero. When I conveyed this message to Dumont, he thought it demeaning to make an overture to Micombero, saying, "He knows where I live."

Needless to say, this was a tense period. Our communications officer collapsed from overwork. The closed-in communications room was so hot that the communicators were working in their underwear. Given the security situation, we could not hire someone to install a window, so it took a new can-do administrative officer to take a sledgehammer and knock out an air vent. We destroyed our sensitive papers and maintained regular contact with our consulate in Elizabethville (now Lubumbashi) in the Congo, which was prepared to evacuate us by boat on Lake Tanganyika if necessary.

The British had hoped to be expelled from Burundi for Christmas because of their policy in Southern Rhodesia (now Zimbabwe). We beat them to it but not for Christmas. Early in January 1966, a confidential source reported that there had been a Politburo meeting (as it was called) in which it was decided that the ambassador, myself, our public affairs officer, and a very knowledgeable local Greek employee would be given twenty-four hours to leave the country, rather than the forty-eight hours given the Chinese. To try to confirm this report, I quickly went to a meeting place of the radicals at the Paguidas Bar and was fortunate to see a Congolese who worked in the Foreign Ministry. He had found the Tutsis xenophobic and wanted to go home, but having been on the losing side of a civil war, he had approached me to help him return home. We exchanged a number of drinks, and he finally said, drunkenly: "You know, Monsieur Buchanan, I don't hate Americans, but if the foreign minister tells me to write a note saying . . . what can I do?" I went home and told my wife to cancel the party we were planning to give the next day and start packing.

At 8 a.m. on Monday, January 8, 1966, the chief of protocol and the chief of police appeared at the Embassy. I allowed only the chief of protocol to go up to see the ambassador. So it was that on January 8 we four "spies" drove out ironically between the Jeeps we had given to the Hutu police, leaving the real spy in the Embassy as chargé. I like to think that the radical minister of education insisted on coming over and shaking my hand at the airport because he appreciated my effort at dialogue with a Tutsi government obsessed by its fear of the Hutu majority.

After packing and closing up the residence, Nan and Barbara followed us by American military transport from Bukavu to Kinshasa. They were accompanied by one of our officers, Jay Katzen, who had collected a menagerie of thirty-two animals ranging from a chimp to a one-armed chicken. They had to get rid of the jealous chimp, Antoinette, in Kinshasa when she threatened their first-born child.

# 8

# A Different Hot Spot: Gabon

The fact that Personnel had reassigned me as deputy chief of mission to Gabon within twenty-four hours of my expulsion from Burundi suggested that Gabon must be some hellhole, difficult to fill. In fact, I enjoyed Gabon. Libreville, the capital, like Freetown in Sierra Leone, was named for the slaves released there from captured slave ships when France joined England in combating the slave trade. In 1839 and 1841 the French negotiated treaties with the two Mpongwe clans who controlled the opposite banks of the Gabon River. Under the treaties, the Mpongwe kings agreed to end the slave trade and accept French sovereignty.

A country consisting largely of virgin jungle, on the equator between Congo- Brazzaville and Cameroon, Gabon was hot, hot, hot. After one dance, I would have to go outside and wring out my shirt. The air- conditioner in my bedroom was the only one that semiworked, so I left the living room doors open for air. There were the usual bugs. At times the dining room table seemed to be moving with thousands of miniscule spiders. Huge crabs in my small garden ate my plants and, on one occasion, chased the wives of French officers off my croquet pitch. At breakfast the next day, my Chadian cook pointed to a huge crab emerging from the guest bathroom. Previous ambassadors had felt that we should live like the French without screens on the windows, with the result that the legs of female staff members were red and swollen from mosquito bites. My ambassador, David Bane, quickly remedied the situation.

## French Paranoia

Gabon was the "jewel in the crown" of Francophone Africa. A French paratroop battalion may still stand ready at the airport to intervene, as it did in the case of a coup against President Léon M'Ba. Gabon is fantastically rich in Okoume mahogany, its original source of wealth, but also in iron ore, manganese, and uranium for the French *force de frappe*, onshore and offshore oil, and indeed anything of value.

The French were paranoid about any threat to their control of Gabon. I learned how paranoid they were early on. When Ambassador Bane introduced me to the foreign minister, whose wife was French, he told the ambassador that Monsieur Gally would like to see me at my earliest convenience. Bane had no idea who Gally was. It was soon clear that the muscular young man who came to see me was the head of French counterintelligence. I suddenly realized that the French assumed that, because I had come from Moscow and been expelled from Burundi, I must be a CIA station chief arriving to weaken French rule in Gabon. There was nothing I could have said, I suspect, that would have convinced them to the contrary. In fact, since Gabon had no relations with communist countries, CIA had no post there.

I was able to play on French conviction that I was a CIA agent to take advantage of the bitter relations between Gally the counterintelligence officer, and the rather sinister hunchback representative in Africa of Jacques Foccard, the Elysée's "Mr. Africa." They each tried to play on American fears of Cuba, exaggerating the number of Cubans in neighboring Congo.

French perceptions that I was CIA proved to be a problem, however. President Omar Bongo had requested American foreign aid to balance the overwhelming French presence, but when Ambassador Bane explained to him that we did not believe wealthy Gabon needed our aid, he turned to me as note-taker and embarrassingly asked: "And what does Mr. Buchanan have to say?" Bongo had learned from his friend, Congolese President Mobutu, that CIA Station Chief Larry Devlin regularly provided funds to Mobutu, and he assumed that I would do the same for him. He was obviously encouraged in his misperception by French intelligence.

On one occasion while I was chargé, I received an unprecedented call from the Presidency telling me to be at the airport in half an hour. When Bongo arrived there, he exploded from his limousine without waiting for anyone to open the door and ordered me into the VIP lounge like a child: *"Vas la dedans"*(Go inside). With blazing eyes, he waved a piece of paper in my face that turned out to be an obviously French-written intelligence report intended to destroy American influence in Gabon, which may indeed have reflected the real fears of paranoid French security officers.

Bongo was told that the American doctor, Fergus Pope, at Schweitzer's hospital in Lambaréné, was at the center of a spider-web organization, run by me, designed to destabilize Gabon. Dr. Pope had allegedly told the staff at the hospital that he would return when there was a revolution. I quickly told Bongo that there was some tragic misunderstanding and proceeded to rent a plane to fly down to Lambaréné to clear up the situation. Fortunately, I caught Dr. Pope as he was off on a trip and asked him to repeat what he allegedly said, suspecting that the problem might be Pope's poor French. What he claimed to have said could have been distorted by someone interested in doing so.

What most angered me was the revelation that Dr. Pope had taken it upon himself, without consulting the embassy, to set up a vaccination program in each of the towns where we had Peace Corps volunteers (PCVs). And he decided that the most honest Gabonese to assist the Peace Corps were Protestants, without realizing that the most feared rival of Bongo was the Protestant former minister of education. It was easy to see how French intelligence could fabricate a convincing theory of how Dr. Pope was organizing a network of Protestants and PCVs to undermine Bongo, under my direction of course.

The fact that, unlike the French ambassador, I could not expel a national in twenty-four hours for embarrassing relations with the local government only further convinced Bongo that Pope was an important cog in my subversive scheme. We breathed a sign of relief when Pope finally left Gabon. Later, when Bongo was convinced that we had no CIA station in Gabon, he was reportedly quite put out, asking why Gabon did not merit a station.

**Schweitzer Hospital**

We had problems with Dr. Pope for some time. A former Air Force pilot in Korea who reportedly had refused to shoot down enemy planes, he was one of the good doctors recruited by Dr. Schweitzer by appealing to their idealism. To retain these doctors, Schweitzer would not only hold their passports hostage but would also appeal to the ambition of each doctor, implying that he was the one selected to replace him. But Dr. Pope had a larger ambition—to reform the way Schweitzer had run Lambaréné and in the process to challenge the French medical system. To the old staff at Lambarene, whatever the good doctor Schweitzer had done was sacrosanct. Accordingly, they resented the fact that Pope had allowed the staff to set up its own soccer team, when Schweitzer had derided soccer as a game for children, saying, "If they want exercise, they should plant a tree." Their resentment of Pope's *lèse majesté* presumably found a sympathetic audience with French counterintelligence operatives.

Meanwhile, Gabonese intellectuals and foreigners were inclined to trash Lambaréné as a dirty, badly run hospital, but they overlooked Schweitzer's basic concern: to provide a hospital environment reassuring to patients and their families. For in Gabon there was a tradition of voodoo, poison, and cannibalism among rival tribes. Hence, each hospital room had space outside where each family could cook for its sick relative. Schweitzer also designed the sleeping quarters for the staff, which, while they provided little privacy, became the model for the schools that the Peace Corps was building in Gabon. Taking a Gabonese hurt in an automobile accident to the more modern but filthy Libreville hospital only convinced me that, if anything happened to us, we should go to Lambaréné, where the doctors were better. Nancy used to be told by the nurses at the Libreville hospital, to whom she was teaching English, that cases of cannibalism were common.

We spent one Easter at Lambaréné with Schweitzer's daughter. After dinner, we visited the leper colony, shaking hands with the pitiful patients. It had rained on our way back and driving on wet, red clay African roads is like driving on soapy glass. Even with four-wheel drive, I had trouble staying on the road. As a result we missed the ferry at the Kongo River. The tom toms were drumming in a nearby village and the mosquitoes were swarming. I managed to find a fisherman who paddled me in his dugout *pirogue* canoe across the kilometer or more of river only to find that there was no one to man the ferry. Our fisherman was so accurate that he hit the same stick in the river going and coming. Back at the car, the fisherman brought a larger *pirogue* and his cheery wife who helped him paddle back across the river. Our German Shepherd seemed as nervous as we were that the *pirogue* might tip over and provide a tasty meal for a crocodile. We spent the night in an inn run by a woman from Marseilles. It was one of the filthiest hotels we ever endured, but her Kongo River crayfish cooked Provençale style were fantastic.

## The Peace Corps

The Peace Corps program of building schools in Gabon (along with teaching English) was if anything too popular, with *préfets* competing to have a school in their district. The Peace Corps philosophy had changed by the time I reached Gabon. The first group of volunteers had been selected for their construction skills, and they quickly built schools, but without any real participation or learning by the Gabonese. When I arrived, the aim was to involve and teach the Gabonese so that they could build their own schools.

There was a problem, however, in areas like Lambaréné, where the local population was accustomed to being paid for work. It was hard for our PCVs to get unpaid volunteers. It was pitiful to see a tiny Gabonese boy bringing water to two PCVs perched alone in the blazing heat on a roof beam. By contrast, in southern Gabon, where the population was less accustomed to being paid for work, a Yale architect PCV taught himself the local language, *Baponou*, and promised three dynamic young local chiefs that he would build them schools if they would help by collecting rocks and other things. As a result, the embassy ended up with three schools for the price of one. The difference was local tradition and local leadership.

One of our favorite volunteers, Jay, the son of psychiatrists, had adopted and totally spoiled a little chimpanzee named Gordon that screamed when left alone. We visited Jay and his chimp at his camp hundreds of miles into the equatorial jungle. There we enjoyed our bucket shower but were put off watching Jay eat off the same plate with the chimp. The trip back down the Ogoué River in our Zodiak, skirting rocks and whirlpools, was spectacular.

En route home to the States, Jay arrived late one night in Paris with a sick Gordon. With his friend's help, they finally persuaded a pediatrician to come over and see their "baby." Unfazed when told what sort of baby was involved, the doctor said, *"Au boulot"* (Let's get to work) and treated the monkey. When Gordon visited our house in Maryland in his black coat, red pants, and pumps, he terrified our African Shepherd and almost broke my leg with a casual slap of his arm. So off it was to the zoo.

## Embassy Life

Aside from monitoring political developments in Gabon, the embassy mainly concentrated on economic developments in this Kuwait of Africa, notably the large American economic investment in iron ore and manganese and the growing sector of oil exploration, on and off shore near Port Gentil. The few local representatives of American companies would complain about the lack of interest shown by their headquarters in the African market, in sharp contrast to active sales activity on the part of the Germans and Japanese. As a result, there was not much consular activity.

But what I remember most vividly involved the deaths of U.S. citizens. The first was a young Jew from a religious family who had drowned in the Ogoué River. Under Gabonese law, reflecting the equatorial climate, anyone who died had to be buried within twenty-four hours. The boy's parents were most insistent that his body be dug up and returned to the States for a proper Jewish burial service. It was a gruesome job for our consular officer.

I was much more personally involved in the second case. At 5 a.m. one morning I received a call that our communications officer had committed suicide. It was hard to believe. The man in question was a bluff, hearty, smiling personality, which he evidently used to cover up a difficult family relationship. My wife had the painful task of telling his little boy that his father was dead. To ship his remains back to the States with his family within 24 hours, he had to be prepared for travel by the funeral parlor by noon, when the only plane was leaving for the United States. I spent most of the morning at the Palace trying to get a document signed authorizing the shipment of the body. With the plane waiting while a religious service was performed over his coffin, I was still at the Palace as a large, uneducated policeman typed out a long form, slowly with one finger, driving me up the wall. We just made it, but ever since I have been suspicious of individuals who seem unusually happy and outgoing, what I term the Pagliacci personality. I felt that as DCM, I should have been more sensitive to the inner pain of a member of our staff.

## Jungle Travel

One of the great pleasures of Gabon was the opportunity to fly deep into the interior of the equatorial jungle to visit either PCVs or some of the great mining sites in which American firms had a large or controlling interest. A frustrated World War II French fighter pilot would say, "Do you want to see an elephant?" and put his Cessna into a deep dive over an elephant seen flapping its great ears with alarm. Or we would fly up the riverbed below the tree level over fishermen paddling their *pirogues,* or above the surf along the spectacular Gabonese beaches.

One of our more memorable trips was to the iron ore–mining site of Belinga, with majority ownership, as I recall, by American Steel. We flew through cumulus storm clouds and arrived in a torrential downpour. A shovelful of earth at the site contained 65 percent iron ore. The IMF turned down a request that it finance a railroad connecting the several hundred kilometers from Libreville to Belinga on the grounds, as I recall, that the global market for iron ore was weak. It failed to take into account the huge reserves of Okoume that would have been opened to exploitation.

My most enjoyable trip was accompanying a group of delightful jazz musicians through the Gabonese interior. The group leader—I think he was called Randy Weston—was almost seven feet tall. Finding him a bed was a problem, and by the end of our tour only eleven keys were still functioning on his electronic piano. Sometimes there would be no electricity in a village. The villagers especially appreciated an old musician from New Orleans who was a terrific drummer.

## Gabonese Beach Life

Ambassador Bane, who was in his fifties, had learned to water-ski in Malta en route home from his previous post in Pakistan. As a result, his first request to me was to persuade the State Department that we needed a boat with twin motors for potential "evacuation purposes." A boat given to us that was originally scheduled for Lake Chad was not seaworthy enough to handle the twelve-kilometer estuary, with its strong current dividing a long spit of land from the mainland. And so we were given a motorboat with twin Mercury engines that constantly needed repairs.

It was almost a ritual on weekends to prepare a picnic lunch and motor across the estuary to the embassy's designated beach property. Water skiing was part of that ritual, which was fine for those who were athletic and had been skiers. One poor new administrative officer, Al Warnecki, transferred from Prague, was so petrified by our innocent discussion of sharks and barracuda that he remained bent over his skis too scared to fall.

To try to find beach property on the mainland for embassy personnel frightened by the estuary, I contacted the mayor's office. We sailed up and down the coast until I saw an attractive beach with lovely sand and palm trees. The problem was the strip of virgin jungle between the beach and the nearest road. With a couple of my "boys," we set about cutting a trail through the jungle. It was not the fear of snakes that deterred us but a strip of swampland. My wife worried that I might be tempted by the mayor's offer of a plot of beach for myself, but it would of course have been illegal.

I had already succeeded in building a beach house across the estuary for our Sunday picnics. Built of logs and thatch, it compared favorably with the beach house of the French Embassy. As a frustrated architect, I derived much satisfaction from the project, which proved quite challenging, having to load our supplies on a large raft that we towed over to the beach. We dug posts into the sand as a breakwater. But what was more unusual, we built a fireplace. In the dry season, when the temperature dropped to 80 degrees, we would be chilly, what with our thinned Equatorial blood, and put on sweaters.

## Religious Experience

Another project was fixing up the Episcopal Church at Baraka. Built of New England pine, with pine pews, it was a reminder that the earliest missionaries to reach Gabon in 1842 were Protestants, not Catholics. The gravestones in the overgrown cemetery we cleared underscored the short life span of the early missionaries in fever-infested Gabon.

Despite all the efforts of the missionaries, Voodoo, with its use of the hallucinatory root *Iboga*, remained an influence in Gabonese life. Quite sophisticated officials told me seriously that with the help of *Iboga* they had been able to talk to their ancestors. A French *coopérant* (a type of Peace Corps volunteer) who used it had to be evacuated to France, so the root had clearly bad side effects. The witch doctors in the villages used *Iboga* in their role as psychiatrists treating the mentally disturbed. Late one night deep in the jungle, I watched as a witchdoctor tried to bring a young woman who had lost her husband out of her deep depression. She was painted white and seemed to be drugged. After dancing and chanting, we all drank from a bowl filled with a mixture of cheap red wine, palm wine, and a sprinkling of sand. I did not stay to see the results of the therapy.

## The French Community

Inevitably in a country where the French ambassador was, in effect, the governor general, much of life revolved around the French community. Ambassador Maurice Delaunay was a highly sophisticated French diplomat who had driven across the United States in his Ford Mustang. Concerts at the French Cultural Center were often interrupted by the squealing of thousands of bats that lived in the dead trees near the Center and blackened the sky when they emerged at sunset.

On the equator, sunset is like pulling down a shade, and it is suddenly dark, happily ending my one-sided tennis matches with Mr. Gally, a category-two player in France. Judo, my other competitive sport in Gabon, was even more one-sided. The building on stilts would tremble as my "partner," a 200-pound butcher from

Lille with a brown belt, would slam me to the deck, to the admonition of our instructor, "A*llez doucement*" (Take it easy). I quickly decided that Judo was not good for "old men." (I was 40).

There were two categories of veterans in Gabon—the early *colons,* who had made their money cutting the huge forest of Okoumé, and the refugees driven out of Vietnam who could not stand the cloistered life of France. Among them were Vietnamese like my neighbor, who liked elephant hunting—a much more dangerous sport in Gabon, hunting in the dense jungle, than in the open savannah grasslands of East Africa. Having been forced to flee Vietnam, many of them strongly supported the American war effort, and more than one Frenchman asked why we did not drop an atom bomb on the communists.

The advantage of serving in a Francophone post in Africa over one colonized by the British is that the local cuisine is infinitely superior. In Gabon we had excellent Vietnamese restaurants. Racial relations were also generally better than in former British colonies. The Gabonese would shrug off a typical racist remark by one of their old *colons;* but if a new arrival from France was obnoxious he risked being expelled in twenty-four hours by the all-powerful French ambassador. Interracial marriages were common, but they had their price if you were a prominent official. The foreign minister was a Fang, and any Fang arriving in Libreville assumed he would find bed and board with "his" minister. The foreign minister and his French wife would accordingly turn the lights off early in their house and pretend that they had gone to bed.

### Gabonese Tribal Politics

Most of what I learned about Gabon came from Minister Pierre Fanguinoveny, who was married to a Swiss lady of Lausanne. They had several children. Historically, the coastal Mpongwe was the most sophisticated tribe in Gabon, selling slaves to the Europeans, and reputedly conversant in several European languages. While Pierre was a Fang, he too was no slouch. He spoke French, English, German, some Russian, and fourteen Gabonese dialects. Originally nomadic cannibals, the Fang were the major tribe in Gabon. Small, black as pitch, Pierre had the presence of someone of royal blood, and indeed it was his father, a Fang chief, who gave Dr. Schweitzer the land on which he built his hospital. Schweitzer was reputedly so impressed by the elaborate funeral of Pierre's father that he asked to have the same funeral for himself. True or apocryphal I don't know. One sensed that Pierre was not an intimate of Bongo, but was treated with respect as an important Fang.

Coming from a small tribe near Franceville, Bongo protected himself politically by ensuring that his cabinet was tribally balanced, even if it meant appointing someone not very competent. He was also skillful at co-opting and neutralizing his political enemies, appointing them as ambassadors abroad rather than having them killed. He always played it safe when he and his powerful wife made their annual shopping trip to the United States by insisting that he be accompanied by his whole cabinet, damn the expense, so that there would be no one left behind to plot against him.

For years Bongo pressured Washington to invite him on an official visit to meet with the president. We finally compromised and arranged in 1977 for a working meeting with President Jimmy Carter, but with no Blair House or official festivities. At the time I was back in the Africa Bureau as office director for Central Africa. The name "Bongo" was obviously a political liability from a publicity standpoint. Unlike one of the first Soviet officials to visit the United States, Shitikov, whom we insisted was pronounced CHITikov, Bongo was Bongo. I was sent up to New York in an executive jet to bring Bongo to Washington. Since Bongo's primary concern was to be seen on Libreville television with President Carter, it was agreed that only the Gabonese press, and not Bongo's cabinet, would be invited to the White House. I was accordingly shocked when my desk officer called me from the airport to say that the cabinet had commandeered all the Gabonese press cars and was en route to the White House. President Carter could not have been more gracious, shaking hands with each Gabonese minister in the Rose Garden, as well as yours truly.

In my time as office director, I was grateful that the chief of protocol was the lady I knew as Shirley Temple. I had been her escort officer on one occasion in Moscow, and as a former ambassador to Ghana she was sensitive to the craving for respect and attention on the part of the African diplomatic corps.

**Recruiting Africanists**

When I finished my tour in Gabon in 1967, I was assigned to the
Naval War College in Newport, Rhode Island. But for family reasons, we needed to remain in Washington. Gabon had been a difficult post for the family for a number of reasons, with Nan spending much time with our children back in the States. The personnel
officer, my friend Alan Lukens, claimed that the only job he could
find in Washington was his own, which would free him to go out
as DCM to Senegal. Having no negotiating leverage from Gabon, I
had accepted, only to be approached by Phil Habib, urging me to
join him in trying to negotiate a peace agreement with Vietnam.
Partly out of loyalty to Alan and partly because of my strong an-
tipathy to our policy in Vietnam, I turned Phil down. To this day I
regret my decision.

I had major reservations about becoming a personnel officer;
remembering names and faces was not my forte. But after a few
months poring over personnel dossiers, I surprised myself and others by how much I remembered.

Persuading FSOs to go to Africa was another story. The excuses
proffered why an officer could not possibly serve in Africa were as
amusing as they were irritating: "I can't remember all those countries"; "My wife could not tolerate the heat"; "How will we educate our children?" and so forth. To make Africa appear inviting, I
filled our offices with the most attractive pictures of embassy housing, along with palm trees. But my major pitch was something I
still strongly believe, that a junior or middle grade officer serving
in a developing area like Africa will be given more responsibility,
meet officials at a higher level, and get more personally involved
in projects helping the local community than he or she would ever
have in prestigious Europe. From the standpoint of influencing the
often young political leaders in Africa, perhaps the ideal FSO (or
CIA operative) was a young, unmarried officer, who would have
no problem frequenting the local bars where cabinet officers often
hung out.

When the personnel officers from the different bureaus met, we
would be competing for the best officers and trying to dump our
"turkeys." It was a rather cold-blooded operation. My goal was to

develop for the Africa Bureau the sort of career corps we had in the Soviet field so that officers who had performed well in a series of hardship posts would automatically qualify for a pleasanter post in Europe or at least North Africa. The European Bureau did not cooperate.

# 9

# Back to Soviet Affairs

In the spring of 1968 I returned to Soviet Affairs as chief of the Bilateral Relations Section of the Soviet Desk (SOV). In time, I moved up to become de facto deputy to our chief, Spike Dubs, whom I had replaced in Moscow. It was a crisis period as the Soviet Politburo debated whether it could afford to allow Czech communist leader Alexander Dubček to follow through with his experiment to establish "Socialism with a human face" in his country. Conservative Czech communists saw themselves losing control of their country. And the Soviets, having decided that the implications were too dangerous, intervened militarily with a façade of East Bloc states, excluding Romania and Yugoslavia. In Washington I was chain-smoking along with four-pack-a-day Spike and decided it was time to stop before it was too late. Meanwhile, our bosses Chip Bohlen and Mac Toon took off to play golf, persuaded that there was nothing that we could do that would deter Moscow.

## Constructing New Embassies

Two issues dominated my two years on the Soviet desk. The first was the issue of embassy housing. Both the Soviets in their building on Sixteenth Street and we on the Ring Road in Moscow badly needed more space. Initially the Soviets tried to buy properties at Bonnie Brae and then Tregaron, but in each case the neighbors said: "No communists in our neighborhood." To get around this problem, we offered the Soviets federal property on Mount Alto. Initially the Soviets were not happy, but soon their intelligence specialists must have pointed out that this site high over Washington was perfect for radio intercepts.

Security concerns bedeviled the whole embassy construction issue. The problem I wrestled with in talking to the Foreign Buildings Operations (FBO) staff was how we could protect ourselves against the installation of "bugs" in our future chancery. I proposed that all panels to be attached to the building be produced offsite in Germany or Scandinavia. But no one insisted that the steel structural beams be constructed offsite, probably for economic reasons. The Soviets took advantage of this, installing a listening device system within the beams that was more sophisticated than anything we had seen before.

The Soviets argued that they were prepared to trust construction of their chancery to a local American construction company and we should do the same and hire a Soviet firm. When I returned to Moscow as political counselor in 1970, the issue was still unresolved. Soviet ambassador to Washington Anatoly Dobrynin was a smooth and persuasive interlocutor; and when Ambassador Jacob Beam seemed inclined to agree with him, I intervened. Whereupon, in typical Dobrynin fashion, he said to Beam, "I guess we know where the holdup is with this project."

Eventually construction went ahead on both sides with Sea-Bees, the Navy's Construction Battalions, doing much of our work. I was told that the Soviets protected themselves against any bugs we might have installed in their new building by burning out the inside of each office. FBO spent millions tearing our new building apart, reportedly to see how the new Soviet listening system operated. We should simply have decided that the new building could not be made secure and locate offices there that did not have security requirements—like USIA, Commerce, and the Department of Agriculture, which had selected expensive office space across Moscow as far from the Embassy as possible.

## Different Ways to 'Bug' Us

I had personal experience of Soviet bugging. In 1963 we found that the bricks of our Chancery on Tchaikovsky Prospekt were hollow and hid a bamboo-tube transmitting device. The best tube was behind the radiator in my office, where, as one of my newspaper contacts noted, I used to brief him on Soviet developments. My neighbor Bob German, who was peripherally involved in the Penkovsky case, found a bug in his mattress. I recall telling a new officer in my living room in the Embassy that we should try to get in touch with a particular Soviet, and within minutes I received a call from the gentleman. And there was always the perhaps apocryphal story about the Marines having a grand time on New Year's Eve, commenting on "those poor bastards" who had to spend the night listening to them. Whereupon the phone rang, and there was a loud pop of a champagne bottle being opened.

Leningrad had its share of bugging problems. When our Consulate Building at Number 15 Petra Lavrova (now Furshtadskaya) was being readied for occupancy, I was told that the staff staying at the Astoria Hotel used to go over every morning and dig the bugs out of the wet cement. When I was consul general, my French and German colleagues warned that they had found bugs in their office ceilings. But it was only after I had left Leningrad that our security experts finally found our bugs. One could protect oneself against radio intercepts by being careful of what one said and going into our supposedly bug-proof "box" for serious professional discussion or airing sensitive personal problems. It was, however, unrealistic to expect all the staff to wait until they got into the box to let off steam.

Trying to deter the Soviets by countermeasures could be dangerous. I was told about a West German debugging expert who transmitted a high frequency signal over the Soviet bug, perhaps destroying the hearing of the person on the other end. Within days he was almost killed with a clandestine injection of mustard gas.

A new threat emerged with the microwaves aimed at the Chancery, notably the ninth floor, where the ambassador and key officers worked. Originally called TUMS, I learned about them by chance on my first tour. On my second tour, officers were told about the microwaves and given an option (which no one took) to reject the assignment. There were many theories about the purpose of the microwaves, including the disruption of our own intelligence efforts, but I do not know if we have a definitive answer.

When Ambassador Walter Stoessel died of leukemia, his wife Mary Ann was convinced that the microwaves caused it. A number of wives also claimed that the breast cancer rate was unusually high. But a study by Cornell University concluded that the cancer rate at the Embassy was not higher than the national average.

## Our Leningrad Consulate

The other issue that preoccupied me in EUR/SOV was that of opening consulates in Leningrad and San Francisco on a reciprocal basis. For us it was the first consular post to open outside Moscow since we were forced to close Vladivostok in 1948. The question of the consular district was however quite complex. The Leningrad consular district was to run from the Baltic States up to Murmansk and Archangel. But we continued to refuse to recognize what we considered the Soviets' 1939 occupation of the Baltic States. A formula was finally worked out whereby the capital cities of Riga, Tallinn, and Vilnius fell within the Leningrad consular district, with our embassy in Moscow responsible for the Baltic hinterland.

In 1969 I was detached from a congressional delegation to Moscow and sent up to Leningrad, accompanied by the embassy security officer, to look for property to rent. Foreigners could not buy Russian land in Soviet times, and it remains a sensitive issue. I was surprised to learn that we had never owned, only rented, property before the revolution; and the location and size of our chanceries depended on the pocketbook of our emissary.

Consulting the 1914 Baedeker, I found that we were renting at the time No. 11 Khalturina, behind the Hermitage. Its original name Millionay seemed very appropriate for our capitalist state. But on investigation we found that asking for No. 11 would have involved evicting too many families. My next choice was Petra Lavrova, which has reverted to its prerevolutionary name of Furshtadskaya. The Soviets offered us a building at No. 15, almost opposite our original chancery at the time of the revolution, No. 24.

In the 1930s George Kennan went up to Leningrad to see if there was anything of value in our old chancery, for which Norway had been the custodian after our departure in 1920. Culver Gleysteen, who had opened our consulate in Leningrad and was my predecessor there, was persuaded by the Soviets that the Wedding Palace at the end of our street had been our original chancery. But a photo Kennan took of the original building in the 1930s showed that it was the two-story building with a large archway across the street from our new consulate. The Foreign Ministry agent accompanying Kennan had insisted that nothing of interest remained in the building. But Kennan ducked up a stairway to the second floor and found a glass door behind which lay the original chancery library, with its set of State Department regulations apparently untouched.

## Travel Controls

My Bilateral Section on the Soviet desk was responsible for supervising travel controls. The one leverage we had in trying to expand our own travel in Russia had been to handle Russian embassy personnel travel on a reciprocal basis. It was difficult to persuade other Americans of the wisdom of our tit-for-tat policy, particularly when we would cancel at the last minute the travel of a Soviet officer scheduled to be their principal speaker. The Soviet desk would consult with the intelligence community to see whether there were any reasons to refuse travel. Our intelligence community was not always so professional. No one objected when the Soviets rented a summer house on the Chesapeake Bay within monitoring range of U.S. Naval test facilities, nor when they rented a summer house near one of our secret underground retreats near Front Royal. I was also convinced that the house the Soviets chose for their Consulate General in San Francisco, high above the bay, was not accidental. But no one raised any objections.

## Counselor in Moscow

In July 1970 I returned to Moscow as political counselor, living in a representational apartment in the Chancery cloistered behind two burly Soviet militiamen guarding the Embassy's entrance. We arrived in time for the traditional July 4 reception in the garden of the ambassador's residence at Spaso House. A sugar baron's manor, Spaso House is redolent of a very different Moscow than the one to which I was now assigned.

Anyone who has read FSO Charles Thayer's delightful book *Bears in Caviar* will remember how our prewar ambassadors, William Bullitt and Joseph Davies, lived in a style to which few professional ambassadors could aspire. It tells how Mrs. Bullitt borrowed birds from the zoo to go with her décor of live trees. When the birds somehow got out of their cages, it took several days to capture them in the four-story ballroom, where they were relieving themselves on the antique furniture. And there was the reception where Marshall Budyenny of Russian Civil War fame insisted on picking up in his arms a young bear, the symbol of Russia, but the bear was not house-trained, forcing the marshal to retire in what had been his white uniform. Or when Bullitt and Thayer set out to teach polo to Budyenny's Cossacks, who regarded their mallets as weapons with which to demolish the opposition. It was an era when a naive political ambassador like Davies could be sending dispatches home justifying Stalin's show trials, a time when the bachelors of the embassy were cavorting with Russian ballerinas without being afraid of the embassy security officer, when Marjorie Merriweather Post, married to Ambassador Davies, bought the basis of her collection of Russian decorative art now on display at Hillwood (her former home in Washington, D.C.). And we often forget that nationwide travel controls were not imposed in the USSR until 1939.

Yes, the life of foreign diplomats in Moscow changed radically after the 1930s, but not for the better. There was a marked increase in harassment, particularly for NATO military. We were told how two of our attachés were drugged, and how another NATO military attaché was almost beaten to death in Gorky Park (in this case there was almost immediate retaliation in his NATO capital). I particularly like the story told by two NATO military officers about how they got even with their surveillants by camping out in a warm tent in the woods on a particularly cold night, leaving their "goons" to freeze in their car until they came and cried uncle.

I was reminded how little had changed when I saw an old acquaintance from Washington, Soviet first secretary Victor Mikhailov, at the July 4 party. He could not wait, he said, to show us his new apartment. The invitation was never followed up. I can just imagine a Foreign Ministry security officer telling Viktor, "Are you out of your mind?" Sad, because Viktor had wanted to reciprocate our hospitality in Washington.

When I was on the Soviet desk I had thought that Russians

cloistered in their fortress embassy should be able to experience a bit of American hospitality. Nancy and I therefore asked three couples from the Soviet Embassy, including the Mikhailovs, for a picnic in the Potomac on Goat Island, owned by a friend. But the water was rather high, and I could see that the Russians were nervous, so we started our cocktails on the Virginia shore before paddling out to the island for lunch and a swim. It was then that I realized that the Russians could barely swim, and I had visions of headlines reading "American Diplomat Drowns Russians." But all went off well. Viktor had simply wanted to thank us for what had been a memorable afternoon.

Soviet security measures ruled our lives. If we wanted to travel, we had to give the Soviets forty-eight hours' notice and a detailed itinerary, from which it was very hard to deviate. When we were told that "for reasons of a temporary nature" we could not travel to a previously open area, or that there was an unpleasant incident like someone having the tires pierced on his car, we would automatically check to see if this might be retaliation for something that had happened in the States.

Burly militiamen at the chancery gate intimidated Soviet visitors to the Embassy; but there were a number of instances when Soviet dissidents tried to get into the Embassy by either rushing the gate or scaling the Embassy wall. One night I found myself negotiating with a group of Pentecostals from Siberia whose children had not made it through the gate when their parents rushed it, and they wanted their children back. It struck me that some might have been inspired by an earlier incident during my first tour to Moscow when some thirty-six bearded Siberians made it into the Embassy and left only when the Soviets promised that they would not be punished for their action. I asked the much less impressive Pentecostals about this event, and they told me, to my surprise, that the local party chief in Siberia had been reprimanded for his treatment of his constituents, which had forced them to seek sanctuary in our embassy.

The KGB colonel with whom I had negotiated all night refused to return the children until the Pentecostals left the Embassy. He told me angrily to get rid of them. "*Vam nye nada, nam nye nada*" (You don't need them, we don't need them). They finally left in the morning. After I left Moscow, I heard of another group of Pentecostals who got into the Embassy and refused to leave and were fed for months by sympathetic staff.

## Reporting on China

On this tour I had a thoroughly professional staff who were regional experts with responsibility for following events in their own area through the Soviet press and contacts with other diplomats, Soviet journalists, and academics. To cover Asia, I was blessed with Stapleton Roy, who spoke Mongolian as well as Russian and went on to be our ambassador in Indonesia and China. We were handicapped in one area of special interest because we did not have relations with Communist China. Our British colleagues liked to tell us what we were missing in the way of dinners at the Chinese Embassy. We wished to monitor relations between the Soviet Union and China, which had openly deteriorated since 1960.

Aside from their ideological rivalry for leadership of the world communist movement, an underlying racial tension gripped the Russians, particularly in the Far East, concerning the danger of Chinese expansion into the thinly populated areas of Siberia and the Maritime Provinces, at one time part of China. Jingoistic Russian politicians exploited the "yellow peril" theme for their own political agenda. In 1959, just before I returned to Moscow, Russia and China fought a brief war over control of Damansky Island in the Amur River. Reflections of this friction appeared in Moscow, where Soviet students were appalled by the diligence of their Chinese colleagues at Lumumba University. There were periodic, carefully calibrated Soviet demonstrations against our Embassy, usually focused on Vietnam. But on one occasion the Chinese students decided to show their bourgeois Soviet allies how real proletarians demonstrate and proceeded to throw rocks and bottles of ink at the Embassy, causing damage that the Soviets were required to repair. Soon we enjoyed the sight of Soviet militiamen beating up the Chinese students with obvious satisfaction.

## The Politics of Détente

Another issue I followed closely was an apparent debate in the theoretical press over the advantages and dangers of pursuing a policy of détente with the West. A note of commendation from Hal Sonnenfeldt, by now on the National Security Council, regarding a dispatch that I wrote on the subject reassured me that at least someone in Washington was reading our output.

The issue was especially pertinent because preparations were under way for a summit meeting to be held in Moscow between President Nixon and General Secretary Brezhnev. Henry Kissinger was so paranoid about premature leaks of confidential plans that he would spend three days negotiating secretly with the Soviets at their Lenin Hills guesthouse and condescend to tell Ambassador Jacob Beam before leaving as little as necessary regarding his meetings. Malcolm "Mac" Toon, later to be ambassador to Moscow, was furious that Jake Beam had not resigned on the spot; but Jake was too much of a gentleman of the old school, whereas Toon was a feisty Scot.

To add insult to injury, Kissinger would use Soviet interpreters rather than one of our own, thereby depriving himself of a trustworthy translation of what the Soviets might say, both in negotiations but also in chatting within their delegation. The importance of conveying accurately both the tone and the substance of conversations was illustrated for me when Senators Henry Bellmon and Hubert Humphrey met with Prime Minister Kosygin and I was note-taker. We rather hoped that both sides would clear the air with some blunt talk, but Viktor Sukhadryev, a fantastic interpreter in both American and British English, carefully edited what each side said, rounding off any sharp edges.

Like so many advance parties for presidential visits, the hubris and arrogance of the Nixon party had to be seen to be believed. They treated the ambassador like some flunky in his own home and had most of our political officers counting the number of steps that Nixon would have to make walking in the Kremlin. I was only peripherally privy to the unseemly way that Kissinger treated Gerard C. Smith, who was our expert arms control negotiator, to ensure that he and not Smith was given full credit for the Anti Ballistic Missile Treaty. But the tension was palpable.

For me the most interesting part of the 1972 Summit was the banquet in the Kremlin. Before being ushered in to dinner in the great Saint Georges Hall, we invited guests were introduced to the principals. If an alien from Mars were asked who was the communist and who was the capitalist, he would have singled out Nixon as the communist, walking like a frozen body inside a glass capsule, while red-faced, hearty Brezhnev could have been the mayor of any town in the United States running for office.

While I did not like either Nixon or Kissinger, I was impressed by their strategy for dealing with the Soviets, using the opening of relations with China as leverage in negotiations with Brezhnev. I also agreed with the Lilliputian concept of tying up the Soviets with a whole web of agreements going way beyond arms control. At the last minute I even found myself checking the text of a Treaty on the Protection of the Polar Bear in the Arctic. An enthusiastic Brezhnev was reported to have told Secretary of State Kissinger that if we had any good sense we should preserve the détente between us and thereby ensure a global condominium.

## The Jewish Question

Kissinger's strategy of combining carrots and sticks in dealing with Moscow was conceptually sound, but the U.S. Congress upset this calculation. Responding to the Jewish lobby, it took away the economic carrot by passing the Jackson-Vanik bill, which tied economic concessions to Moscow to the level of Jewish emigration from the Soviet Union. Later, for balance, it added Pentecostals and Ukrainian Autocephalists, who it was presumed had suffered persecution and warranted refugee status like the Jews.

Jackson-Vanik was not an unmitigated blessing for the Jewish population of Russia, at least for those that were well integrated into Russian society. A friend of ours, a well-known writer of children's stories, said that up to the time of the turmoil caused by Jackson-Vanik she had never really thought of herself as Jewish. The fact that Jews were given an opportunity to emigrate to America but Russians were not aggravated latent anti-Semitism. There was already jealousy because the Jewish population as a whole was more prosperous than the Russian.

The Russians responded sullenly to outside pressure. A huge amount of time at both the embassy and later the consulate general in Leningrad was spent pressing the Soviets to grant exit visas to Jewish families on our lists and trying to find out what had happened to individual Jews reported by the Jewish grapevine to be in prison. Jews in Russia would phone relatives in New York, who would contact the State Department, who would contact us. And I would send an officer down to the synagogue; if the information proved solid, it might provide the basis for another representation to the Foreign Ministry. Some of the Jews attending services in the synagogue were devout old parishioners with their black and white outer garments, familiar with the Torah. But many young Jews knew less about the service than I did, but would gather enthusiastically after the service to sing Israeli songs on the steps of the synagogue. An annual issue taken up with the Foreign Ministry was the importation of matzo for the high holidays.

The Soviet authorities did not know how to handle Jewish human rights activists. Many Jewish visitors, especially from England, would arrive with lists of Jewish contacts in Russia, providing the Jews with a reassuring sense of solidarity and with reading material not available in Russia and returning with information on what was happening in the Russian Jewish community. The Soviets even went so far as to rough up some of the visitors. Even the officer who was my main contact with the Jewish dissident community when I was consul general in Leningrad was knocked down by some "hooligans" on his way to one of his sources. The officer was Daniel Fried, a former nanny in Moscow who went on to become assistant secretary of state for European Affairs.

Jackson-Vanik has remained a thorn in the side of U.S.-Russian relations even though Jewish emigration is no longer an issue. Many of the Jews who immigrated to Israel have returned with an Israeli passport to use their experience to do business in Russia. Successive administrations in Washington have promised Moscow that Congress would repeal Jackson-Vanik, only to find repeal linked in the Congress to some totally unrelated issue like the export of chickens to Russia.

## The Human Rights Movement

We had always hoped to see the Jewish human rights movement and the largely Russian movement join forces to increase their leverage, but they continued to go their own ways. The Russian human rights movement was particularly effective with the regular distribution of dissident writings, *samizdat* (self-publication) and *tamizdat* (publication abroad), that were read and passed from hand to hand. Khrushchev lifted the lid off liberal expression a crack by allowing the publication of Aleksandr Solzhenitsyn's *One Day in the Life of Ivan Denisovich*, exposing life in the Gulag. The authorities tried hard to put the cork back into the bottle and prevent distribution of both *samizdat* and *tamizdat* but to no avail. The KGB tried different ways to suppress the infection by locking up biologist Zhores Medvedev in a lunatic asylum, by forcing Solzhenitsyn to emigrate to the United States along with our future poet laureate Joseph Brodsky, and by exiling Andrei Sakharov to the closed city of Gorky.

I regret that I never had the opportunity to meet these luminaries of the Russian human rights movement. About the only things they had in common were their hatred of Stalin and the pale imitations of him that emerged after his death. But they had very different visions of the good society. Medvedev remained a convinced communist and follower of Lenin. Sakharov's views evolved from talk about the convergence of the capitalist and socialist systems to a firm belief in democracy, while Solzhenitsyn remained until the end the moral voice of Russian Orthodoxy and Russian national tradition. Nobel Prizes awarded to Boris Pasternak for his *Doctor Zhivago* in 1958 (which he was forced to reject) and later to Solzhenitsyn and Sakharov served to encourage and provide legitimacy to Russian dissenters.

But the main boost to the human rights movement came from the 1975 Helsinki Accords, the Final Act of the Conference on Security and Cooperation in Europe. The Accords linked "cooperation in humanitarian and other fields" with the territorial settlement of postwar boundaries Moscow sought. The dissidents were quick to use the Helsinki Accords to pressure their government on issues of human rights, as Western negotiators had hoped.

Given the difficulty of holding frank conversations with Soviet officials, or even average Soviet citizens, journalists reporting on Russia naturally tended to treat the dissidents and artists as the only real voice of Russia. There were, in fact, many faces of Russia that we normally did not see. Our surprise that the system could have produced a Gorbachev and a Shevardnadze reflected analysts' difficulty in looking behind the façade of a unified party line to evaluate the importance of any hints of differences that emerged from a careful reading of the party press.

## Religion

While religious feelings contributed to disillusionment with the Soviet system, the Russian Orthodox Church as an institution was too subservient and infiltrated by the KGB to play any positive role in the evolution of Russia. Religion was a hollow shell of what it had been. In addition to the synagogue, there were forty-five active churches in Moscow in my day, survivors of the ten thousand churches Khrushchev destroyed throughout the Soviet Union. Still, the potential for a religious revival that President Vladimir Putin ironically promoted was always present in the beauty of the Russian Orthodox service.

We found the Eastern service most moving, with the elderly *babushky*, scarves around their heads, praying on their knees on the hard stone floor (in Russian Orthodox churches, the congregation stands throughout the service), and the glorious singing, the rich apparel of the priests as they emerged from the holy doors in the iconostasis, the sea of lit candles as we walked three times around the church at Easter service. It was easy to understand why the Slavs from Kiev visiting Constantinople in the tenth century were stunned by the beauty of the Byzantine service and opted to become Orthodox Christians. A trickle of monks and nuns have returned to their monasteries and nunneries, and the ability of rural communities to rebuild their churches has been remarkable.

The church of the Old Believers in Moscow is impressive, with its twin choirs that sing across the church to each other, reading the music from huge panels on which the musical notes were inscribed. In the seventeenth century the conservative Russian peasantry refused to accept Patriarch Nikon's reform of the liturgy, accusing the patriarch and all his priests of being children of Satan. And many of the fundamentalist Old Believers, or *Raskolniki*, have refused to this day to have anything to do with the Patriarch or his priests.

After my retirement I visited Niokolayevsk, an Old Believer community on the Kenai Peninsula in Alaska, which had split on theological grounds because the more liberal part of the community had imported an Old Believer priest from Romania. The fundamentalists denounced the "heretics" and moved out, accusing their schismatic neighbors of having burned down their separate log church. Everyone was speaking Russian; the girls wore Russian blouses, while the fundamentalist boys riding all-terrain vehicles near their new home were dressed like Leo Tolstoy. When I asked one of them why they had left Nikolayevsk, he looked at me like an idiot and said, "*Oni poposvstiy I muy bez popovtsy,*" which a seminarian on Kodiak Island explained to me meant, "They are pro- and we are anti-priest." It was just another sad example of the infighting that has historically torn the Russian émigrés apart.

The Baptists are another religious community with a long history, having moved to Russia at the end of the nineteenth century. The Soviets persecuted an illegal underground Baptist movement, which I was told maintained clandestine relations with the legal movement, whose representative in Central Asia I have already mentioned. In Moscow we attended a legal Baptist wedding of an English student marrying a Russian girl. It was an excruciatingly long service, as the pastor spoke at length about the responsibilities of marriage, and the English student had had to memorize long passages in Russian. When I later attended a Baptist service in Leningrad, it was clear that the Baptists attracted younger workers and had a large youth choir. As the honored guest, I was expected to kiss all fifteen bearded priests in their receiving line. What we do for our country!

Protestant services were regularly held alternately in the British and American embassies with a Church of England pastor stationed in Finland. The Catholics attended their own church. Normally there were no problems, but I was called down one day to the Foreign Ministry where the deputy head of the USA section, one

Fedoseyev, had trouble keeping a straight face as he protested the behavior of a member of Congress who had walked down Gorky Street handing out Bibles.

## Relations with the American Section

Fedoseyev was the most pleasant member of the very competent USA section of the Foreign Ministry. One wondered how he survived in a section headed by as tough and argumentative a diplomat as Viktor Komplektov, unless the rumor was true that he was KGB. He could ill-conceal his satisfaction when reporting to us that an exploding bus tire outside the Foreign Ministry had shattered Komplektov's leg.

Fedoseyev's wife was reportedly brought up in the Bronx. Their sophistication in English came in handy when we invited the USA section to a buffet supper and a movie. It was a trauma. Our Soviet maid encouraged our very well-mannered German Shepherd to eat the five chickens Nan had prepared for dinner, requiring an emergency fallback. And I had planned to show a well-publicized movie, when I was warned that it reinforced every Russian prejudice about America. As a result I showed David Niven in *The Statue*, a risqué comedy about a public relations executive in London who tried to discover the identity of the model his wife had used for a commissioned statue—or more precisely for a particular portion of his anatomy. The Russians clustered around Fedoseyev's wife in the intermission as she explained some of the finer points of the movie. They must have enjoyed the movie, because Indian diplomats approached me later and said, "We hear that you have a *very* good movie.

## Relations with Artists

Some of our most enjoyable evenings were those spent with dissident artists. The chairs and crockery might be broken, but the hospitality was always overwhelming. The widow of a well-known Lithuanian artist, Valius, used to give lectures in her basement apartment to young Russians about modern art and her husband's works, forbidden topics. We bought an abstract painting by Valius of a lonely figure called "Nostalgia," looking at the lights of Moscow in the distance.

One of the most valued contacts in the artistic community was a Greek employee of the Canadian Embassy, Costakis, who had amassed a huge collection of avant-garde paintings, both from well-known artists like Malevich and the new dissidents. Costakis had decided it was time for him to return home to Greece, but what to do with his collection? Eventually he negotiated a deal with the minister of culture (and reputed mistress of Khrushchev), Yekaterina Furtseva, under which the Tretyakov Museum would retain perhaps two-thirds of the collection and identify Costakis as the source of the paintings when they were put on display. I saw that they abided by this agreement when I visited the museum some years later.

Another well-known source of dissident painting was the American journalist Ed Stevens, or to be more exact his very tough and enterprising wife Nina, who reputedly had been previously married to a high Soviet official of the Ministry of Interior (NKVD) and apparently maintained her contacts. When forced to leave their beautiful three-story log house (the type that Khrushchev destroyed on a large scale to make way for modern buildings), they were given a handsome house and garden decorated with beautiful furniture, some dating from Peter the Great, not far from the Embassy. Nina was the businesswoman. Sadly, Ed, who won a Pulitzer Prize in 1948 for his reporting on Russia, had replaced journalism with booze.

## Skiing Mount Elbrus

As difficult as it often was to maintain contact with official Russians, we were able to use the years of détente to advantage. Unbeknownst to each other, the Austrian ambassador and I had both been turned down during our earlier tours in Moscow on our requests to ski Mount Elbrus, at 18,000 feet the highest mountain in Europe. In 1972 the Austrian request was again rejected, but presumably as a gesture of goodwill in the year of the Nixon-Brezhnev Summit, the Soviets authorized our trip.

We were put up in a rather attractive A-frame hotel, though with the usual plumbing problems. The architects had conceived of the grand idea of hooking the water system in the hotel to the nearby mineral Narzan Springs—like having Evian water emerge from the faucet. But the Russian pipes could not handle the corrosive minerals in the water, and they began to leak, allowing sewage to enter the drinking water and making a German tourist delegation very sick.

The young (ostensible) ski instructor had obviously been given instructions to organize a party for us with all the golden youth of Moscow and Leningrad who were staying in the deluxe, reportedly military sanitaria nearby—an attractive bunch of young people. Talking to our host, Nan expressed interest in local Caucasian art, and the very responsive young man promptly set up a meeting that night with a local artist, an old man who was probably Cherkes, as we were in the Karachaevo-Cherkass Oblast. The artist proved to be quite deaf and violently anti-Russian, which he made no effort to conceal. The next day he presented us with a brass bas-relief of a panther rampant, which he said was the symbol of his people. This was the last time we saw either the artist or our young host.

It was apparently decided that we needed to be put into the more experienced hands of an important party official in charge of constructing both the lift and the hotels on Elbrus. He said his name was Viktor. He had been on the Soviet downhill ski team and spent two years as an engineer at the Soviet Antarctic base in Mirny. Assuming again that the facts were not fiction, he worked for the Ministry of Medium Machine Building. His beautiful red-headed wife Olga was a heating engineer, but Nan soon found that she was more interested in fashion magazines like *Elle* than in engineering. Viktor was certainly a stronger skier than either of us, and he took us up the crude lift to ski down through the slushy spring snow. We did not try to go up to a youth hostel located, Viktor said, at 15,000 feet.

One evening after dinner with Viktor, we returned quite late to find the doors to Hotel Elbrus locked. Our hosts then insisted that we return to their apartment and sleep in their large double bed. When we woke the following morning we found the living room crowded with people who had also spent the night. It appeared that housing was at a premium and friends of friends of theirs from Moscow would arrive unannounced. As the guests of honor, we were treated at breakfast to what some Russians at least considered a delicacy, a small slab of uncooked white lard.

The sad part of any prolonged relationship with a Russian in those Cold War days was the paranoia that accompanied it. There was nothing that we did together that was not suspect, like our night in Viktor's apartment or even the time they brought us wild rhododendron from the Caucasus. They clearly had a hunting license to frequent us. I was accordingly surprised when Viktor became upset when I passed him on the street a bag of the latest fashion magazines for Olga. He must have known that we were being photographed. Nevertheless, we had good times together, exchanging turns as host at a local restaurant.

Many years later, we learned that Olga had divorced Viktor, as in most Russian divorces because he drank too much, and had married a most unlikely American from Oklahoma. Now a Russian businesswoman, she had met my old Foreign Service colleague Bob German at a conference, and he gave her our address.

To find an excuse for returning to Russia after retirement, I persuaded the mayor of Jackson Hole in 1984 that it would be great if Jackson were to become a twin city with Teberda-Dombai, a nature reserve and spectacular alpine ski resort in the Russian Caucasus. I renewed contact with Olga, who arranged my sister-city adventure to Dombai, on which my sailing FSO friend Alan Logan and my doctor friend Bill Brewer accompanied me.

When we arrived in Moscow, our tour guide, a surgeon and alpinist, had arranged meetings with the Ministry of Economy. I tried quickly to disabuse our hosts of their obvious hope that this "rich American" was coming to help develop Dombai as a ski resort.

Our host introduced us to a small, bespectacled gentleman who, he claimed, had been a schoolmate. Igor quickly explained that he was retired from the KGB and had just toured the United States with a CIA-sponsored group, lecturing about the importance of intelligence in the post-Communist era. "As I recall Tom," he said, "your station chief in Leningrad was X." He was right, of course.

The Russians were understandably curious what an old cold warrior was doing in a sensitive national minority region like the Caucasus, and accordingly inserted Igor into our party.

Caucasian hospitality lived up to its reputation, and I had my first hangover since college. Once again when I was the guest of honor, I found myself served with that delicacy, a slab of raw lard. I noticed that my KGB companion skillfully hid his lard under his potatoes.

The local Dombai hospital greatly appreciated the medical equipment I brought from Jackson. The mayor of Dombai was quite receptive to the idea of sistering with an American alpine resort. Our group found the scenery breathtaking, particularly when we had to land our helicopter for repairs, cutting short a skiiing expedition for some German tourists.

I was enthusiastic about the idea of helping the local economy by selling the beautiful wool sweaters that the babushkas spread out on the snow for us. Sadly, by the time I returned to Jackson, my supporter, Jackson's mayor, had been fired, and I was too frank, I fear, in describing to the local Rotary Club how we were stranded up in the air for over two hours when the antique ski lift broke down. Jackson had its own development problems, without taking on a rundown ski resort in Russia.

## Soviet Professional Contacts

As illustrated in the case of Viktor Mikhailov, it was not easy to develop more than a formal relationship with Soviet officials. And even when dealing with ostensibly nongovernment officials, one knew that with rare and often suspicious exceptions one was dealing with officials encouraged to deal with specific foreigners. They could still be of interest and were often highly sophisticated individuals. Whether they would meet with us was always a barometer of our relations. On one occasion, when relations were rather tense, I invited Averill Harriman to lunch along with two Soviet officials. One of the Soviets got the message and did not show up. The other one arrived, realized his mistake, but could not leave and remained, visibly sweating.

Among the journalists, I enjoyed very much Viktor Matveyev of *Izvestya*, a shrewd observer of the international scene. I particularly remember a conversation we had on one occasion about the conflict in the Horn of Africa, when we joked about how we kept exchanging client states, Ethiopia for Somalia.

The ubiquitous Viktor Louis, married to a former nanny at the British Embassy, was clearly a KGB informant used to pick up gossip in the diplomatic community or feel out the situation in countries with which Moscow did not have good relations. A poseur, he was given a fancy estate in fashionable Peredelkino, where he kept his Mercedes, a ski-lift, and every form of gadgetry bought in the United States Asked why he had built a sauna when he had no intention of using it, he replied, "So you would ask." He had a global map with pins showing the countries he had visited and those that refused him a visa.

The USA-Canada Institute headed by Georgiy Arbatov was a major contact on all kinds of issues. Arbatov had a very competent staff, some of whom seemed more trustworthy than their leader.

On the one time that I visited IMEMO (the Institute of International Relations), roughly equivalent to Brookings, with a prestigious staff, I was amused by the comment of its director that "of course" none of their women researchers were in positions of authority. On one occasion traveling to Novosibirsk in Siberia, I contacted with difficulty at Academgorodok (academic city), the son of former ambassador to the United States Mikhail Menshikov. He was one of a number of younger social scientists who had left IMEMO to work in the freer, less hierarchical atmosphere of this new academic campus. There he was doing a comparative economic study of how different societies respond to different economic inputs, a study, he noted, driven by the hard conclusions of figures rather than ideology. With obvious irritation, the younger Menshevik finally had to ask his KGB watchdog to allow us to walk alone "around the campus," which he compared to Reston, Virginia.

Traveling by train to the Baltic states, William Luers and I were put in the compartment of a sailor and his girlfriend. The Russians make no effort to keep the sexes apart on their trains. Our companions were amply supplied with spirits, and we had some Bourbon, which came in handy when a KGB officer insisted on joining our conversation. He downed a tumbler full of Bourbon Russian-style, bottoms up. When I encouraged him to have a second glass, he hesitated, but pride prevailed. He downed another tumblerful, turned green, and rushed away not to be seen again.

## Diplomatic Life

As the reader can imagine, diplomatic social life was rather frenetic, and Moscow was a hub of self-corroborating rumors whirling around the city. The Latin diplomats clearly were the ones who had the most fun, dancing into the wee hours. We were closest to my opposite numbers among the Canadians, British, and French who were well-informed, delightful people and, over time, good friends. There were also German diplomats who knew Moscow well, having helped in some cases to rebuild it as POWs during the war. A Finnish diplomat named Kaarpinen had an impressive knowledge of Soviet society and politics from years of study and residence. So we were surrounded by stimulating colleagues.

The usual number of amusing incidents lightened the rather somber mood of Moscow. Unfortunately we were not at the French Embassy for the dinner party at which, we were told, a waiter lost control of the trout on his platter. One slid off the platter and down the décolleté bosom of one of the ladies, only to be retrieved in a flash by the waiter and put back on the platter. Now that is *savoir faire!* And then there was the time one diplomat went to the Bolshoi Opera and tried to check two hats at the *garde-robe*, explaining that he had picked up the second hat for a friend. The little old lady in charge flatly refused, saying no man can have two heads.

## Family Life in Moscow

Conditions have radically improved for families in Moscow today. At our first small apartment on Leninsky Prospekt, the washing machine took up much of the narrow entrance hall. Campbell used to play in the large communal playground and skating rink for the inhabitants of the surrounding apartments. He would come home boasting of having traded chewing gum for Tsarist coins— not strictly legal. Or he would return from visiting a Russian playmate and tell us that Igor had a much bigger apartment than we did—Igor's parents were doubtless *nomenklaturas*. When we held a children's party and invited Campbell's Russian playmates, they would arrive, usually much better dressed than the American boys, and obviously nervous about having to walk past the militiaman stationed at our entrance.

Boys will be boys, and the lack of play area for those living in the Chancery meant that they sometimes got into trouble. On one occasion they lit a fire in the elevator shaft. Only after I left Moscow did my wife dare tell me that Campbell and other boys would climb the lampposts and pull down the red flags during Soviet holidays.

We adults had more outlets for our energy. For the staff, broom-ball, played in ordinary shoes on ice, was the primary sport until the ambassador banned it because of the high casualty rate. DCM Walter Stoessel and I would occasionally play hockey. I was not on our team when, with the help of an Air Canada former professional hockey player and some Scandinavian ringers, we beat the TASS news agency team. When we went on the ice for a return match, our Soviet goalie (we did not have enough players) said: "TASS hell, that's the second line of Dynamo" (a Soviet professional team). We did not win that one.

For the families of all the diplomatic corps in Moscow, there was an Anglo-American embassy-run school, which Campbell attended. But five sons of the army attaché attended a Soviet school, with the eldest son graduating first in his class. Soviet security officials refused to let him take his prize, which was a camping trip down the Volga River. On one occasion some Russian bullies reduced the youngest American to tears, accusing his father as I recall of being a murderer in Vietnam. He was not a popular boy, but his schoolmates knew a political provocation when they saw it and beat up the bullies.

Even some ideological aspects of Soviet education could be quite funny. One mother who used to help a very nice Russian kindergarten teacher by playing the piano for special events discovered that the kindergarten was celebrating Red Army Day and her son, as the smallest boy in the class, was told by the Soviet officer in charge to climb to the top of the human pyramid and wave the red flag. Mother had her revenge, playing "America the Beautiful" as the class filed out.

Our closest friends in the diplomatic corps in Moscow were the Canadian minister Pierre Trottier and his statuesque British wife Barbara. A great linguist, Pierre was also one of Canada's most distinguished poets. He went on to become Canadian ambassador to Peru and then to UNESCO, sadly ending his days as an alcoholic. But we would listen with envy as Barbara and Pierre told us about the farm they had picked up for a song in France in Provence, while we sipped the *petit rosé* wine they had brought back to Moscow. This was the beginning of a magical twenty-three years in our lives worth some elaboration.

# PHOTO GALLERY 2

U.S. Embassy Bujumbura, Burundi, 1965

Embassy gives a truck to Burundi farm cooperative, 1966

# L'ambassadeur des Etats-Unis expulsé du Burundi

Bujumbura, 10 janvier (A. F. P.).

Le ministère des Affaires étrangères du Burundi, agissant au nom du gouvernement, a signifié, lundi matin, à M. Donald Dumont, ambassadeur des Etats-Unis, d'avoir à quitter le royaume dans les 24 heures. L'avis d'expulsion touche également MM. Thomas Buchanan, conseiller de l'ambassade, et Nicholas Milroy, directeur du Centre culturel américain à Bujumbura.

Un employé de nationalité hellénique de l'ambassade des Etats-Unis a également été déclaré pernonnage « non grata » et prié de quitter le Burundi dans les 48 heures.

News coverage of DCM Buchanan being expelled from Burundi with Ambassador Donald Dumont, January 1966

Receiving a Meritorious
Honor award, 1967

DCM Buchanan with
the U.S. Ambassador
to Gabon, David Bane,
1967

U.S. Embassy
Libreville,
Gabon

Episcopal Church, built in 1834, at Baraka, Gabon

Touring the interior of Gabon with Randy Weston's
band, 1967

Peace Corps Volunteer Jay Leven and Gordon, his chimpanzee,
Gabon, 1967

Embassy Gabon wins second prize in Libreville floral competition,
1967

Office Director for Central Africa Buchanan visits Emperor Bokassa
during visit to the Central African Republic, 1976

With the U.S. ambassador to the Central African Republic,
Anthony Quainton, and CAR Minister of Foreign Affairs Francke

In signed photo with Foreign Minister Francke

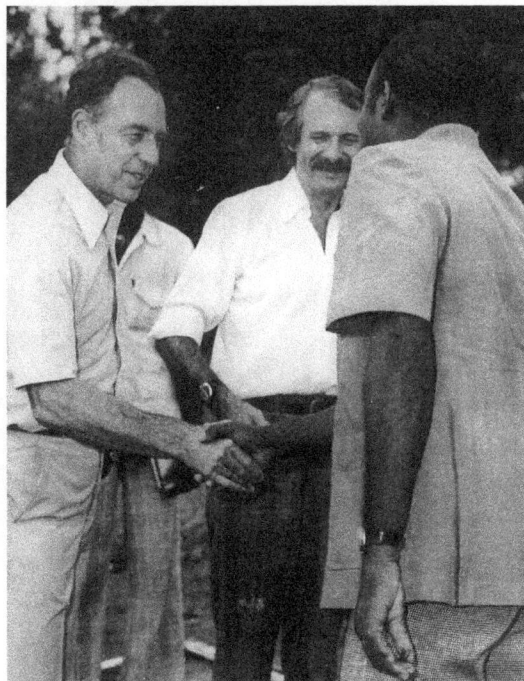

With U.S. Chargé William Swing in Central African Republic, 1976

U.S. Chancery on Tchaikovsky Prospect, Moscow, U.S.S.R.

At a performance of the Bolshoi ballet in honor of Nixon-
Brezhnev Summit meeting, 1972

November 7 parade on Red Square, Moscow, 1972

DCM Buchanan gives departing ambassador to Norway Philip Crowe a farewell gift, 1973

With Crowe's successor as U.S. ambassador to Norway Thomas Byrne, 1974

Consul General's residence in Leningrad, U.S.S.R., 1977

Consul General Buchanan (at left) as Master of Ceremonies at Embassy Moscow farewell party for Ambassador Malcolm Toon, 1979

# 10

# Our House in Provence

We were encouraged by the Trottier's tales, a small inheritance, and some leave to look around Provence for ourselves, just for fun we told ourselves, with no intention of buying any property. But once we had found Saumane de Vaucluse, not far from the Trottier farm, and looked across the valley at its twelfth-century church and the chateau of the infamous uncle of the even more infamous Marquis de Sade, we were hooked. The little hilltop village of Saumane de Vaucluse was twenty-five minutes from Avignon, an hour from Aix en Provence, shopping distance from Isle sur la Sorgue—known for its *brocante,* or market for secondhand furniture—and within sight of Menerbes, where Peter Mayle wrote his amusing book *Toujours Provence.*

We found a small three-story village house in ruins, with a garden and grotto on the hillside behind that overlooked the tile roofs of the village. We learned only later that the village *pissoir* (urinal) went with the property—a source of amusement among our friends.

Madame Martin, our chain-smoking real estate agent, quickly discovered that the property was owned by seven members of the Fautrero family, all of whom hated each other and insisted on using different notary publics. The prospects of getting the house did not look good, and we decided to look elsewhere, beginning with my childhood stomping ground, the Riviera. In Cannes a real estate agent was amused at the idea of our buying property for what we said we could afford and suggested that we look in the mountain area of the Var, near the Gorges du Verdon. And so we exchanged the enchanting perfume of thyme and rosemary of Provence for the bracing mountain air of Les Basses Alpes (the Low Alps).

## Our Village in the Var

The area had its own charm, as we approached through the fields of lavender the little town of Moustiers Ste. Marie, famous for its pottery. Beyond Moustiers we found Aiguines, a small hamlet with a tile-roofed château high on the hillside above a new reservoir, the Lac de St. Croix, with the submerged village of St. Croix visible under the waters. A real estate agent had suggested that we look at Aiguines, and sure enough we found another half-ruined property with a garden behind, overlooking the chateau, for 12,500 francs or ça $2,800. We bought the property in October 1973 for around the same price as our ruin in Saumane and took steps to gift it to our son Campbell to balance the gift of land given to daughter Barbara in Jackson Hole. The Buchanans had second thoughts, however, when Campbell made clear that he was not interested in any property far away in Europe. And the mayor of Aiguines informed us in September 1976 that we needed "immediately" to take care of the tiles and rocks that threatened to fall down on passersby and should take steps to consolidate the remaining structure. The mason proposed by the original owner of the property to do the work gave us an estimate of 14,864 francs, or more than the cost of the property. We loved swimming in the Gorges du Verdon but were also sickened by the pollution from the campers along the lakeshore. And so in 1977 we agreed with alacrity to sell our Aiguines property to a neighbor.

## Saumane de Vaucluse

By now we were focused on developing our property in Saumane de Vaucluse. To our surprise, we had found when we returned from buying our property in Aiguines that Mme. Martin had bullied the Fautreros into selling their ruin of a house, and Mr. Grandvillemain his small adjoining plot. The sale took place in October 1973 with our friend, the pharmacist Jeanette Reboul, serving as an intermediary, since we were by now in Norway. In her wonderful Provençal accent, Mme. Martin said, "I would prefer to drink all the Mediterranean than to do this job again."

# 11

# A Norwegian Surprise

It was Bob German, then political officer in Oslo, Norway, who opened the next chapter in my bureaucratic life after Moscow. Bob called me one day from Oslo saying that Ambassador Philip Crowe had received my name from Walter Stoessel (an old friend who became ambassador to the Soviet Union and Germany and under secretary for political affairs). Crowe wanted to interview me for the post of deputy chief of mission, which had been vacant for six months, since John Ausland married a Norwegian pharmacist and retired in Norway. I was so startled I was speechless. Eastern Europe, yes, but Norway? What did I know about Norway? Well, on June 10, 1973, having now been appointed, the Buchanans flew into Oslo on a rare sunny day when Norwegians sat face to the sun and young girls sunbathed on the palace grounds.

Having been paneled without being interviewed by Crowe, I was sent to Washington for the brief DCM course, returning in early July. Meanwhile, Nan had served in effect as the official hostess for the ambassador, who had been accepted as ambassador to Denmark but was determined to host the traditional July 4 party before leaving Oslo. To do so, he pushed up our Independence Day to June 26. Nan was amazed that the ambassador did not recognize the prime minister of Norway, Trygve Bratteli. But then Crowe was unique. Formerly in the OSS (his favorite picture of himself was with some bare-chested thugs from Southeast Asia), he said he had paid $5,000 to the Republican Party and, voila, his diplomatic career took off. An arch-conservative, he told me that the only Norwegians worth dealing with were the wealthy, conservative shipowners, with their salmon fishing rights, and that I should quickly join the Conservative Club—all this in a country brought into NATO by the Labor Party. No wonder Crowe did not recognize the prime minister.

I had my problems when I returned to post as chargé. A man of huge ego, Crowe insisted that he should sign off on my cables to the department, even though he was already ambassador to Denmark. I informed Joan Clark, our wonderful administrative officer in the European Bureau of my little problem, and she finally said, "Oh, let him do it."

I was in a state of shock after my first working day in Oslo. Bob German had introduced me to Kjell Vibe, the real brains and executive head of the Foreign Ministry. Reaching for a subject on which I had some knowledge, I asked Vibe about Norway's complex relations with its huge neighbor, the Soviet Union. Vibe proceeded to lay out the Norwegian position, including differences of opinion within the ministry, with a frankness he would not have encountered in the State Department. In stark contrast, a couple of days before leaving Moscow, I had with difficulty arranged to meet the head, as I recall, of the Southeast Asian Division in the Soviet Foreign Ministry. In Hollywood caricature fashion, he told me in answer to a question that if I had read *Pravda*'s article on the subject the previous day, I would know the Soviet position. We continued our friendship with Vibe and his wife Beatrice when he became Norway's most distinguished ambassador to Washington.

My next shock was the visit to the labor union headquarters. I got off on the wrong foot, opening the door for a woman who was thoroughly irritated by this bourgeois behavior. The goal of Norwegian Labor was, and probably remains, a very social democratic one, providing support to everyone from the newlywed, the families needing daycare centers, stipends for ambitious students, and so on. A more radical protocommunist party challenged Labor on the left.

I was fortunate to see an enviable example of an election in this truly democratic society. The Norwegian public was apparently looking for a change from Labor's long time in office, and the race was tight, with the Conservative and Center parties challenging Labor's rule. Instead of the carefully managed debate of American elections, with little opportunity to challenge a candidate, the leaders of the thirteen or so parties in contention met around a table and asked each other questions, all broadcast on television, not accepting evasion for an answer. Any costs connected to the election campaigns were funded by federal taxes. The outcome was a conservative coalition government led by the Center Party. A deciding issue was one familiar to Americans, that of abortion. My colleague in Moscow at the time, the Norwegian counselor Petter Svennevig, as the only Center Party diplomat with any seniority, was shocked to find himself suddenly promoted to state secretary, the third-ranking member of the Foreign Ministry.

When I paid my courtesy call on Prime Minister Bratteli, I discussed with him the major issue at the time in our relations, that of oil production in the North Sea. With the OPEC boycott of oil sales to the West, Norway was under great pressure, particularly from Washington, to show its Western solidarity by increasing its output of North Sea oil. Bratteli answered me with a bit of history. He recalled that he grew up in a little seaport town that lived on its production of whale oil. When the Greenland whales were killed off, his town died. Bratteli made it very clear that as long as he was prime minister, Norway would be conservative in the exploitation of its oil reserves.

Not knowing how much traveling I would be able to do after the new ambassador arrived, I made a point of visiting as much of Norway as I could while I was still chargé. First Nan and I went to Stavanger, the oil capital of Norway, where we met with the mayor, who asked us what we wanted to do in his town. He suggested that we might like to sail on a replica of the vessel that took the first fifty-two Norwegian emigrants to America in 1834. (Because the vessel was only supposed to carry twenty-nine, the passengers and crew were put in jail on arrival in New York).

Another shock. We had been told that the Norwegians were rather withdrawn people, as difficult to know in their own way as the Russians. Well, as soon as we went aboard, we were slopping down Norwegian cognac, Aquavit, with the usual herring and boiled potatoes. Hospitality was so great that the crew lost their chart overboard and it had to be recovered with a rake. We were dropped off at a port along the way with little memory of how we got there.

It was in Stavanger where an American student at the local high school rented a greenhouse, where he claimed to be raising blue and red corn. To the horror of the Norwegians, who had little experience of our drug culture, they found he was growing marijuana.

Our next trip was to Northern Norway. We first visited our USIA couple in Tromso, north of the Arctic Circle, where it is dark half of the year. My wife said I was a bit paternalistic toward the very pretty blond wife. She had her revenge when they were reassigned. She gave me the lengthy computer study she had made of the Aurora Borealis at the Institute in Tromso, of which I did not understand a word. I had learned in the meantime that, when stationed in Latin America, she had taught herself Spanish so that she could give lessons in computer science to local cabinet members.

From Tromso we went to Bodo, famous for the impregnable underground submarine pens built by the Nazis. The U-2 pilot Francis Gary Powers had been scheduled to land in Bodo when he was shot down over Russia. In Kirkennes on the Soviet border, we talked to the Norwegian officer in charge who described sharing fishing on the border river with his Soviet opposite number.

The Soviet town of Nikolayevsk across from Kirkennes was the site of a tragic story in U.S.-Soviet relations. It was here that a young, mentally challenged American man tried to walk to Nikolayevsk by a shortcut and was arrested as a spy. With the Soviets trying to make an exchange to get back one of their spies, they sentenced the American to eighteen months in a forced labor camp, where he would remain until some deal could be worked out. Somehow he was murdered on the way to the camp, or as the Soviets claimed committed suicide. The FSO who had tried to help the young man had been traumatized.

Finmark, or northern Norway, has a romantic appeal. It is the land of the reindeer herds tended by the Lapps. The emptiness of the land is one of its attractions. We knew one Norwegian lady, a great skier, who was skiing across solitary Finmark when she suddenly saw a lone man skiing toward her. She greeted him with the traditional *Hola*. He glared at her for having broken the pristine stillness of the North.

The Norwegian military were always concerned during the Cold War by the very lack of population in the north, which made northern Norway so vulnerable to Soviet attack. With the risk that the lure of high-paying jobs in the oilfields off Stavanger would further depopulate the north, the government took steps to subsidize the region. "Bridges to nowhere" linking small islands with a minuscule population to the mainland were one result.

If Trondheim on the coast of central Norway was the traditional center where the kings of Norway used to be crowned, Bergen to the south was perhaps the most attractive of Norwegian cities. Its historic role as a member of the Hanseatic League was reflected in its well-preserved architecture. It was the home of Norway's most famous composer, Edvard Grieg. When we later attended Bergen's famous music festival, we listened to a pianist playing Grieg's folk music, and then to a musician playing the identical music on the Hardanger fiddle, not an attractive instrument.

I was delighted to learn from a family memoir that I too had Norwegian roots, like many in our embassy. A Rosenkranz from Bergen, perhaps a younger son in a society of primogeniture, arrived in New York in the 1600s. My cousin in Chicago, Rosenkranz Baldwin, was a direct descendant.

With the arrival of Ambassador Tom Byrne, formerly a representative of the AFL/CIO, his wife Peggy, and seven daughters, the embassy was fully staffed. Our own residence was an interesting Bauhaus-style building built by a shipping magnate, with a two-story living room for his art collection, difficult to decorate but great for a Christmas tree. I quickly started Norwegian lessons, if only to pronounce our street name, Hoffschef Lovenskioges Vei. My trip to the hinterland had persuaded me that even though the inhabitants of Oslo spoke fluent English, this was not the case in the countryside. Anti-German feeling was still very strong in Norway, and my teacher would get very angry when I Germanized a sentence by putting the verb at the end.

Much of our work at the Embassy was helping the ambassador to prepare for his weekly Saturday breakfast with Foreign Minister Knut Frydenlund. Two issues predominated. On one, Norway's production of oil, Byrne was unable to make much progress, for the reasons discussed earlier. We decided that Henry Kissinger might have more success, since Frydenlund had been one of his students at Harvard, and the two were scheduled to meet before the UN General Assembly. But when Frydenlund returned, he told Ambassador Byrne, "Tom, what was all that talk of oil about? When I finally asked Henry, he brushed off the issue as the 'product of the bureaucracy.'" This was a typical Kissinger ploy to avoid a confrontation by putting the blame on "the bureaucrats."

The other issue was NATO and its preparations for a possible war with the Soviet Union. At the NATO base at Kolsås, plans were made for periodic maneuvers and for emplacement of materiel for an emergency. On a NATO naval visit, the flagship, an American carrier, perhaps the SS *Independence*, negotiated the tricky Oslofjord with just inches to spare. (It was in this same fjord, when the Germans invaded Norway in 1940, that a torpedo from an ancient coastal battery sank the German heavy cruiser *Blucher*, with much of the German advance party aboard.) Cocktails before dinner were on a Canadian vessel, since no alcohol is allowed on U.S. naval vessels.

One area where Ambassador Byrne was successful was in selling the F-16 fighter plane to the Norwegians. Our most serious competitor had been the Swedish Viggen fighter.

There was a certain Mediterranean work ethic in Norway, with few officials outside of Kjell Vibe to be found in the Foreign Ministry after 3 p.m. But when they were there, they were most efficient. This left the Americans in the Embassy ample time for recreation, notably skiing the icy trails on the five-kilometer course on Holmenkollen Hill above Oslo. At the beginning it was humiliating to be passed on the uphill stretch by some little old Norwegian lady. On one occasion we were asked out for a ski dinner, which we concluded by skiing down through the trees in the dark. In the summer months, the Holmenkollen was covered with Norwegians looking for mushrooms. There was a volunteer mushroom expert at the foot of the *trik* (trolley) stop at the bottom of the hill to check whether you had any poisonous mushrooms. It was not unusual to see King Haakon riding the trolley like any other Norwegian citizen.

Two trips with Norwegian friends were particularly memorable. Commissioned to find thirteen perfect Christmas trees for a Canadian Embassy benefit ball, we scoured the woods above Tyre Fjord on the property of "Big Bear." Almost every tree was defective in some aspect. Back in our Norwegian friends' comfortable log house, we warmed up with Aquavit and then a lovely wood sauna. We were told that the ladies were in stitches watching their men with their pink bottoms scurrying from the sauna across the snow down to the hole cut in the ice in the lake.

On another trip we visited Norwegian friends on the fjord south of Oslo, where the inlets were carpeted with huge mussels. This idyllic spot suddenly became a site of death when a small plane whose pilot, apparently looking at the girls below, stalled and dived headlong onto the rocks. With the help of some teenagers, I was able to extricate the pilot, hanging upside down in the cockpit. Getting help was not easy, but a military helicopter eventually took the pilot and his accompanying son to Oslo, where the son survived as a vegetable, but the father died. I relived that scene with nightmares for days.

Some weeks later we sailed with these same Norwegian friends in their little sailboat and their four golden retrievers. We missed the African shepherd we had brought to Norway from Moscow, where she had been recovering from distemper. Our wonderful Moscow lady veterinarian was distressed that we had to leave before Mweeza, Kirundi for "beautiful," was fully recovered. And in fact four months of required quarantine in Norway were enough to kill her.

Our friends urged us to stop moping about Mweeza and told us the Norwegians were looking for foster parents for the golden retrievers being trained as guide dogs. And so it was that we acquired Nicholas, a small blond ball of fluff, a tiny golden retriever with an impressive pedigree, selected by personality to be a guide dog. He was not to play with balls but to learn to sit at our feet in the car and travel on elevators. Nicholas announced his presence by emptying all the wastepaper baskets in the house and strewing my underwear around the garden. We ended up paying the equivalent of three golden retriever bitches that the kennel could breed for permission to take Nicholas with us when I was reassigned to Washington. After 15 years, we buried Nicholas in a rainstorm, mixed with our tears, at his favorite site looking out at the Potomac.

## Challenge of the North: Svalbard

I was distressed to be reassigned to Washington in November 1975 just when my long campaign to alert Washington to the problems that Norway faced in the north with its large neighbor had resulted in an agreement to discuss the issue with the Norwegians. Typically, it appeared that Kissinger discounted my many cables on the subject until a German diplomat I had briefed convinced him there was a problem.

Norway's main problem involved Soviet behavior on Svalbard (Spitsbergen) and in the Barents Sea. Norway had been awarded sovereignty over Svalbard by the 1920 Treaty of Paris on behalf of some forty signatories to the treaty. This was an historic area of conflict dating back to the competition for whale oil in the seventeenth and eighteenth centuries. Bloody battles between Dutch and British whalers ended only with the extermination of the Greenland whale. It was from Ny Alesund, which became a research center, that the great Norwegian explorer Roald Amundsen flew to try to rescue the crew of an Italian airship wrecked near the North Pole, one of many such ill-fated efforts to reach the Pole by sled and balloon. Longyearbyen, the capital of Svalbard, which is some 600 miles from the North Pole, was named after an American, John Munro Longyear, whose Artic Coal Company began business there in 1906.

The Russians were among a number of nations mining for coal before the Treaty of Paris. In the 1930s, the Russians were still extracting coal for their industries in Murmansk, but after World War II the coal mines at Barentsberg and Pyramiden had lost their economic importance to Murmansk. They provided a pretext, however, for the Soviets to maintain a larger population on Svalbard than the sovereign power, Norway. Having seen the Nazis use their weather station on Svalbard in the war to help cut Allied supplies being shipped to Murmansk, Moscow was determined to maintain control of the archipelago after the war.

While Moscow formally recognized Norwegian sovereignty over Svalbard, it generally ignored Norwegian efforts to assert their sovereignty by, for example, controlling flights of helicopters or the travel of Soviet mining families to and from Longyearbyen. To reinforce their claims to a disputed fishing area in the Barents Sea beyond Svalbard's 200 nautical miles fisheries protection zone, the Soviets even held naval maneuvers in the area. With the potential for finding oil and gas on Svalbard, but particularly in the Barents Sea, the archipelago has taken on new importance.

I managed to get myself invited to visit Svalbard by the Store Norsk Kulkompanie, a private coal company that ran Svalbard rather like a company town, at that time even printing the currency used on the island. I flew on a private company jet to Longyearbyen where I met with the *sussellman*, the Norwegian governor general of Svalbard, who confirmed the problems Norway was having with the Soviets. Relations between the two communities were spasmodic, with an occasional sporting event. But when the Norwegians beat the Soviets at chess, the Russians promptly imported a couple of "ringers" from Murmansk to ensure this did not happen again.

Upon my arrival at Longyearbyen, I applied to visit the Soviet mine at Barentsburg. I was not surprised when I received no reply, so I decided to at least fly over the site. The only private plane for rent on Svalbard was, as I recall, a biplane, built by its Austrian pilot. Once we had shaken loose of the ice, we flew out over the fjords at a time when the sun projected shadows from the ice mountains far out onto the fields of white. And there we saw her, a great female polar bear, lumbering to get away from our noisy beast. Barentsburg looked just like its photos, a cluster of unattractive brown buildings. To enable them to say they had not refused entry to a signatory of the Paris Treaty, the Soviets in their usual fashion authorized me to visit Barentsburg the day they knew that I was leaving Svalbard.

There has reportedly been an appreciable increase in activity on Svalbard since I left—tourism, higher education facilities, and some high tech satellite relay stations. While the Soviets have closed down Pyramiden, the output of Svalbard coal increased five times to 2.5 million tons in 2004. Presumably most of this is from the Swedish mine at Svea, which I visited, where the miners had created a replica of a Wild West bar, with posters declaring "Jesse James, Dead or Alive." One of the most original projects on Svalbard is the so-called "doomsday" seed bank, storing seeds of as many world varieties as possible in a great tunnel hollowed out of the rock.

Skiing on Svalbard was a disappointment. The snow had a curious, sticky quality. Travel was by snowmobile, with frequent accidents in the spring, when snowmobilers misjudged the thickness of the ice. My favorite relic of my trip was a large fossil given me that showed a tropical leaf dating back to before the Glacial Age.

# 12

# Return to Africa

When I was told that I would be leaving six months ahead of schedule to return to African Affairs, I quickly called the Personnel Office in Washington to protest the decision. I was told that I could come back on consultation, but there was no point since I had been paneled for the new job the previous week. Case closed. This was at a time when the Foreign Service was making a great effort to accommodate the wishes of new officers.

What I learned was that I was the victim of *force majeure*. L. Paul "Jerry" Bremer, who went on to become ambassador to Iraq (where he made a series of disastrous decisions), was aide to Kissinger. Understandably burned out, he asked to leave, and Henry told him he could have the post of his choice at his grade level. Bremer hesitated between Munich and Oslo and chose Oslo.

Meanwhile, Kissinger, with his usual paranoia, had convinced himself that the Africa Bureau was sabotaging his Angola policy of confrontation with the Soviets. He was doubtless influenced by the effort of Assistant Secretary for African Affairs Nathaniel Davis (before Henry reassigned him to Switzerland) to persuade him not to get involved in what was essentially a tribal conflict in Angola. Two of the tribes, the FNLA under Holden Roberto and UNITA under the ex-Maoist Jonas Savimbi, claimed to be anti-communist. But the MPLA under the mestizo poet Angostino Neto had been receiving low-level Cuban and Soviet aid.

Kissinger, who had been waiting for an opportunity to demonstrate that the United States had not lost its will after the Vietnam fiasco, chose Angola as the place to challenge Moscow. Sensing lack of enthusiasm for his policy among Africanists, Kissinger ordered the new assistant secretary, William Schaufele, to replace the key officers on his staff dealing with Angola. And from the standpoint of Bill Schaufele, an old friend, the fact that I was leaving Norway, perhaps not coincidentally at this time, made me the logical candidate for the position of office director for Central Africa, which at the time included Angola.

### New House in Saumane

My reassignment to Washington was the second bit of dismaying news that I received while in Norway. Before that we had received an apologetic letter from Mr. Favre, the very impressive Paris architect working in Bonnieux whom we had hired, telling us that construction costs in France had skyrocketed and it would now cost around $250,000 to restore our village house. That was out of the question.

But then we received a miracle letter from our young French neighbors, who worked for British Petroleum, saying that they had to move away and wondered if we might be interested in buying their house. They had bought the house in 1973 for 210,000 francs but had run into financial difficulties and, after trying to open a restaurant in Saumane, had gone broke. We were friends, but more important we were *les américains, ça veut dire riches*. We jumped at the idea, even at the price of 380,000 francs or about $84,000, which we raised by selling fifteen acres in Jackson Hole. The house clung to the cliff under the chateau, with two of the rooms having the cliff for walls. A little loggia looked over at the Luberon mountain range toward Menerbes, the site of Peter Mayle's efforts to "fit in" and restore a house with that relaxed Provençal work ethic that irritates the Parisians but which foreigners like us found amusing.

For our first Christmas in 1975, I found a huge Yule log for our large Provençal fireplace, and we sat on a couple of broken chairs and the steep steps leading down into our dining room and fireplace, drinking champagne with our Canadian friends, the Trottiers, and their friends, the Rebouls.

## Angola Fiasco

In effect I was selected to preside over what became increasingly clear was a failed policy in Angola. On December 19, 1975, just before I arrived in Washington, the Congress had passed legislation cutting off all aid to Angola. Ironically, the revenues from the Gulf Oil Company investment in the oilfields of Cabinda, west of Zaire, were helping to fund the Neto government in Luanda.

The Cubans, with their own aspirations for a leadership role in Africa, began to provide military advisers to the Neto government in 1975. CIA responded with its own program centered first around the FNLA and then on UNITA. We even persuaded the South Koreans to send a military mission.

My excellent Angolan desk officer Mike Gallagher and I suspected but could never prove that Kissinger had signaled to South Africa through intelligence channels that we would welcome intervention by South Africa to prevent Angola from being overrun by the Cubans. South Africa was happy to have this pretext to clean out the area in Angola from which fighters from the South West African People's Organization (SWAPO) continued to attack South African–controlled Namibia.

The South African intervention was a military success but a political fiasco. Before their entry, we had defended a viable position supported by most African states, namely the need to install a coalition of right and left parties in Luanda. But the South Africans poisoned the well by discrediting in the eyes of most Africans anyone allied with the hated Apartheid regime. The conflict escalated but with America and its allies increasingly isolated.

We worked late every night drafting cables to all relevant governments urging them to press for a coalition government and the withdrawal of the Cuban forces. Once a week our deputy assistant secretary, Ed Mulcahy, and I would attend the CIA Angola Working Group, where we were briefed on the status of our aid to Savimbi, allegedly from monies and supplies already in the pipeline. Since aid had theoretically been terminated, this was one of the longest pipelines in the world! The impressive young CIA officer who used to brief us, the son of a missionary in the Congo, eventually resigned from CIA and wrote a scathing denunciation of this callous manipulation of Africa by the Great Powers.

Savimbi was increasingly isolated. He died fighting in 2002, the South Africans withdrew, and a majority of countries recognized the government in Luanda. In terms of American objectives, it was a total defeat. But the law of unintended consequences came into play. Kissinger had no real interest in hastening the independence of either Zimbabwe or Namibia, nor concern about Apartheid in South Africa. But by converting a tribal war in Africa into a serious East-West confrontation, he raised the specter of a successful Cuban army extending communist influence in Southern Africa.

It was the unintended consequence of a failed policy that persuaded Washington that it must take the lead in the struggle for Zimbabwe and Namibian independence. The emergence of Mandela as a statesman in South Africa eased that transition for Washington. I was among those urging a change of policy in Southern Africa, but Schaufele, who had to deal with Kissinger and all his complexes, said that an approach to Kissinger was premature. Events rather than the Africa Bureau persuaded Henry to radically change course.

## Zaire

As Washington's major ally in the Angola conflict and a neighbor of Angola, Zaire also suffered the consequences of failure. Separatist sentiment in the rich mining area of the Katanga had always been strong. With the civil war dying down inside Angola, the Katangese gendarmes, a radical paramilitary group, decided to return to the Shaba, with some assistance from the Angolans, who saw a chance at payback. Initially the Shaba invasion was a success until Belgian and French paratroopers stabilized the situation.

From the standpoint of the Africa Bureau, President Mobutu was unfortunately a red flag to Congress, and there was little sympathy for providing him with military aid despite all the efforts of Secretary Schaufele to persuade congressional staffers over a whiskey after work that there was a price to pay for inaction. The vulnerability of Zaire was summed up in a wonderful *Washington Post* cartoon showing Zairian soldiers with a catapult saying, "Here they come . . . give them a blast of cheese on dark rye."

It was easy to say "replace Mobutu," but who had the prestige and skill to hold together that huge, tribally fragmented country, two-thirds the size of the United States? The alternative to a permanent Western military presence to prop up Mobutu was obviously military training of the Zairian army so that Zaire could defend itself. Accordingly I headed a small team, first to Brussels, as the former rulers of the Belgian Congo, and then to Paris, which was interested in strengthening its ties with francophone Africa. In each capital I discussed the military situation in Zaire and the vital need to upgrade the Zairian army. Enthusiasm for aiding Zaire was noticeably lacking, particularly in Brussels. But the Belgians also did not like the idea of the French moving in on their hugely rich former colony. For their own political reasons, Belgium and France both sent military training missions to Zaire, to no great effect.

### Visiting the Parish

Before moving on from the Africa Bureau I arranged to visit "my parish," starting with Gabon. I began the trip on a low note, falling in the dark rushing to get to dinner in Libreville. By the end of the trip I had a badly infected leg. In Bangui in the Central African Republic I discussed with Ambassador Anthony Quainton the booming trade in diamonds, which had spread even to countries like Burundi that had no diamonds. I was introduced to the Emperor Bokassa, who obviously liked Tony and used to invite him to spend the weekends at his forest residence. Bokassa had a reputation for being a connoisseur of human flesh; so I asked Tony if he know what he was eating on those visits.

Back in Bujumbura I was amused and irritated when the head of state Micombero, former minister of defense, arrived unannounced and quite drunk at Ambassador David Mark's residence in midafternoon and could not be persuaded to leave until 10 p.m., two hours after the beginning of a dinner in my honor. Micombero kept insisting that it was the King (Mwami Wmambutsa IV) who had expelled me. He protested too much. Since my time in Burundi the civil war had escalated, with coup and countercoup and some 100,000 Hutus reported killed in the 1970–71 period. After the assassination of successive presidents, a power-sharing agreement was negotiated in 2001 that led to the installation of a former Hutu rebel leader, Pierre Nkuruniza, as president, ending the civil war, one hoped.

In Kigali Ambassador Robert Fritts was so impressed by my story of the couple in Bujumbura in which the husband was the accompanying spouse and running the motor pool that he told his staff of unmarried women to go out and get themselves a useful husband.

### Choice: Ambassador or Consul General

It was again my time to move on, but to where? Knowing my feelings about returning to the Africa Bureau, Bill Schaufele had promised to "take care" of me. One advantage of service in Africa is there is little pressure by political appointees to take jobs from Foreign Service professionals. In my case the bureau offered to support a nomination as ambassador to either Burundi or Gabon, where I had served, or to Guinea. I might have briefly hesitated if the neighboring Cameroon had not gone to an African American female historian married to a former ambassador in the Middle East. (The capital city Yaoundé is in the hills and more temperate than Libreville, and the country is rich in folk art.) I was not even tempted by Burundi, with its memories of massacres. Gabon had been a bad time for our family, and the diplomats in Guinea were confined to the capital city Conakry.

The decision became academic. The post of consul general in Leningrad had opened up, and it had always been one of my dreams to serve in that historic city. So I declined an ambassadorial post, which I was told amazed Bill Schaufele, who was then ambassador to Poland—I suppose because becoming ambassador is the presumed goal of every Foreign Service officer. I must admit that I hoped lightning would strike twice, but it did not.

# 13

# Leningrad

To brush up on my Russian before taking over in Leningrad, I arranged to go to the Army language school in Oberamergau, where the bulk of the teachers were Soviet defectors and the students were expected to understand and write Russian. What a shock it must have been for students on their first time in Moscow to arrive from this idyllic environment of mountains, pine trees, and German cleanliness to depressing Moscow, with its sullen gray masses crowding down the vertiginous escalators to the subway, making the rush hour in New York seem like a stroll in the country.

To take the train from Helsinki, Finland, to Leningrad is a reminder of the heroic Finnish defense of its territory against the Soviet attack in 1939, whose incompetence reflected the cost to Moscow of Stalin's 1937 purges of the most experienced Soviet military officers. A Russian scientist once told us how he woke up in his dacha outside Leningrad during the war to see a Finnish ski patrol in his garden. "They made fools out of us," he said. The Soviets insisted on moving the Finnish border further away from their great industrial center of Leningrad. The new border post at Vyborg, formerly part of Finnish Karelia, proved to be a rundown, depressing fortress town, badly in need of restoration. Aboard the train the KGB border guards had insisted on looking under our beds, and they were equally unwelcoming when we arrived in Vyborg.

## Joan Mondale's Visit

My first task when I arrived in Leningrad was to negotiate a visit by Joan Mondale, the wife of Vice President Walter Mondale. Someone should have convinced her that the vice president's wife cannot make an "informal" visit, particularly if it is the first time.

Carter Brown, at the time director of Washington's National Gallery of Art, had told Mrs. Mondale that she had to get in to see the avant-garde paintings in the basement of the Russian Museum. The city authorities were not pleased. If Mrs. Mondale came to Leningrad, it should be on an official visit. We argued for three days whether this would be an official-unofficial or unofficial-official visit And normally neither foreigners nor Russians were given access to the avant-garde collection. To complicate matters further, Mrs. Mondale, who was a potter, asked to meet with a Leningrad potter. The authorities were nonplussed, and I suspected they would show us pots but not the potter.    What was important was that we were given access to the huge collection of avant-garde paintings hung almost floor to ceiling and covering everyone from Malevich, Chagall, and Goncharova to Kandinsky and Popova. Particularly impressive were the knowledge and love shown by the Russian Museum guide for this forbidden treasure in her care. The visit was topped off by walking up six flights of stairs (the Secret Service would not allow us to use the elevator) to meet a charming couple and their daughter, both of whom claimed that the Picasso-like ceramics in the apartment were theirs and had been shown at the Venice Biennale.

Mrs. Mondale had also been badly briefed. Driving back from the airport, when we reached Victory Square she flatly refused to get out of the car and deposit a flower at the sacred flame to the victims of the blockade, arguing that this gesture would make her visit "official." I finally persuaded her that it was essential that she honor the sacrifice of the Leningrad people. I only wish that she could have seen the twenty-minute film shown in the museum underneath the rather impressive monument of semiabstract figures of soldiers and citizens. If she had seen the pictures of people chopping holes in the ice and dragging their dead on sleds through the snow, she would have had a better appreciation of this "hero city," as it is called, and understood the profound fear of war among most Russians of the older generation.

## Living in Style

If in going to Africa I was going from Park Avenue to Harlem, then going from Africa to Leningrad, I was back on Fifth Avenue. I had been responsible for finding our Consulate building at 15 Petra Lavrova, but it was our first consul general, Culver Gleysteen, and his titled Swedish wife who chose as our residence a somewhat grander building than I would have dared propose. But the Buchanans were grateful for their *largesse d'esprit*.

The Soviets claimed that No. 4 Grodnensky Pereulok, on a short dead-end street, had been a gift by Konstantin, the uncle of the Tsar, to a ballerina friend of his, a common practice among the Russian nobility. Subsequent research raises some questions about this claim. In any event, It was a lovely three-story building, which had been restored by the specialists from the Hermitage Museum. What helped the residence to survive the nine hundred days of the Nazi blockade was said to be that, as a home for small children, it had been partially heated. On my first time inside the residence, I had watched a young woman, who looked as though she had come directly off a collective farm, applying complex molding to the ceiling. The woman in charge of restoration told me that she was not following any early plans of the residence but had recreated a typical interior of a manor house of the 19th century. Gleysteen had exploited the visit of President Nixon to Leningrad to decorate the residence with paintings normally reserved for embassies, including several George Catlins, a naïve Charles Humphreys, a huge abstract painting from the Museum of Modern Art, and others. I lost the battle to hold on to the paintings.

We used the palatial living room, ballroom, and dining room largely for official receptions. We preferred a much smaller suite upstairs with its own small fireplace for eating, chatting, and informal entertainment.

It is hard to insulate a palace. When we had three weeks of minus 45 degrees F, Nan and I would huddle in our woolens around one of our two small fireplaces. One of our consuls said he wished he had a larger refrigerator because it was warmer inside his refrigerator than in the room outside. The Russians were embarrassed to have to admit that two hospitals had to be evacuated because the pipes had frozen. Walking our dog Nicholas, with his little rubber shoes, along the Neva near Smolny Cathedral, threatened by ravenous ravens, was a chilling experience.

## Our Life in Leningrad

When we first arrived, we thought we were blessed with a Soviet chef who had been a chef on the liner *Pushkin* that traveled to Montreal. Sasha was never fazed when faced with a last-minute reception for 150 persons. And he would insist that he and not Madame would vacuum the palace when the women on our staff periodically went on strike. Stupidly, we praised instead of complaining about Sasha "to the walls," so that one morning on the eve of a large delegation he came with tears in his eyes to say that he had a better job. I took him out into a little courtyard away from the "ears" and asked what had happened. "They did not tell me," he said, adding in French, "*C'est la vie.*" It was no coincidence that Sasha's brand new car, for which he had waited for years, was stolen at the same time that he was fired from our residence. The Soviets could be so petty toward those whom they disliked or envied.

Lilya, who resembled a slightly plump Marilyn Monroe, was our very efficient housekeeper and took great pride in the residence. Helping Lilya was Nina, scarred in body and mind by the suffering from the blockade. Nina finally was so rude to Nan that I had to fire her, whereupon Lilya came to me in great distress saying that if Nina went she would be fired too. I thought that they would want some "ears" left in the residence, but Lilya was also fired and in the same petty way was made to work at the German consulate scrubbing floors, returning late at night to her dangerous neighborhood. She should have been sent to teach students at the hotel business school in Leningrad how to serve properly and set a table.

I surprised the Diplomatic Agency that provided us with staff by going to Finland to find a chef. They told me, "But that will be so expensive," when I said that I would never be left dependent on them in the future. In Helsinki we found Jacques, who had been chef to the Finnish ambassador to France and was superb. We have never eaten so well. Indeed, one day a button on my vest exploded across the room, and I had to ask Jacques to ease up on our two *cordon bleu* meals a day. Before she left, Lilya made a pass at Jacques, who was in fact gay. They seemed quite intimate as Lilya tried to get Jacques to marry her and take her out of Russia—perhaps a factor in her being fired.

Some of the most unpleasant incidents involved the supposed "handyman" charged with keeping the Residence functioning. On one memorable occasion we found ourselves with nothing but boiling water in all the pipes for several days, no way to get a drink. Our "handyman" came several times and fiddled, to no effect. I finally insisted on calling a city plumber, who came, took a quick look at our heating plant, which resembled a boiler room on the *Queen Elizabeth*, pointed at one valve, and said, "What idiot turned off that valve?" I must have done something to irritate the "Organs," as the KGB was called.

When we were packing to leave Leningrad and in a great rush, I asked our "handyman," who was supposedly supervising the packing, to be particularly careful with two of my prize, delicate African artifacts, a queen figure from Baule and a Guro mask. "By chance," as the Russians say, these were the two items from our shipment that never arrived—probably taken apart to find out what was so special about them.

I was fortunate in having an experienced staff, among them Dan Fried, a future assistant secretary for Europe and Eurasia. He drove the Soviets mad because, in addition to being my contact with the dissident communities, he was a marathoner who ran 18 kilometers three days a week, putting an extra burden on his "tails."

It was my public affairs officer, Cresencio Arcos (who went on to have a distinguished career in Latin America), who persuaded me to accept as a lecturer on American literature Air Force colonel and Vietnam war veteran John Pratt. From the moment John arrived the Soviets sicked their most knowledgeable expert on American literature from the University of Leningrad on him. John said that dinner with his adviser the first night was like passing his PhD orals. Clearly the Soviets thought John was probably a military intelligence officer, but he passed with flying colors. And he and his adviser became good friends. John spoke no Russian, but English speakers from miles around would come to his lectures, and students would quietly ask him questions about American authors whom they were not supposed to have read.

One night at 11:00 I received a call from my consular officer at the hospital with John, who was thought to have a kink in his colon requiring immediate surgery. In a three-way conversation, our embassy doctor stressed the importance of cleanliness in the operating room, since the patient's insides had to be exposed. Whereupon, my consular officer pointed out that the operating room had dirty blankets serving as curtains. Fortunately, a more senior doctor looked at the x-ray and said that John had what they call the "lucky cancer" of the lower bowel, and there was no immediate need for surgery that night.

Two of my officers were then detailed to drive John to Helsinki that night. But the problem was to open the Soviet-Finnish border. I called the number-two man at our Diplomatic Agency, Kozlovsky, a sophisticated KGB officer who like Putin had served in East Germany. Kozlovsky opened the Soviet side of the frontier in half an hour, but the Finns had all gone to bed. But all's well that ends well, because the Finnish doctors were superb. And John went on to teach American literature at Fort Collins, Colorado.

I had another traumatic experience of having to open the Soviet-Finnish frontier late at night. General Edward Rowney, a member of the SALT II delegation visiting Leningrad, was informed at 11 p.m. that he was needed to attend a briefing on SALT at noon the next day at the White House. I delegated two of my staff to drive him to Helsinki after determining that if he caught the Helsinki flight, it could link up with a Concorde flight from London to Washington. Once again Koslovsky arranged for the Soviet side of the frontier to open, and once again the car had to wait for the Finns to open their side of the border. General Rowney made his appointment at the White House.

My administrative officer Don Hays confirmed what I always believed, that the best training for serving in the Soviet Union and dealing with its corruption, bureaucratic suspicion, and incompetence was to serve in a developing area like Africa. Don came to us from the Ivory Coast. I knew that he took care of key Soviet officials, and I did not ask him any questions. When a highly recommended administrative officer from the European Bureau who ruled by the book replaced Don, we began to run into frequent harassment and problems with Customs. While some of this may have been politically motivated, I suspect the main problem was that key officials resented not receiving their expected gift of whiskey or the Christmas edition of *Playboy*.

In the relatively closed society of Russia, one of the useful sources of information was the weekly lecture at the Znanye (knowledge) Society. I would spend two to three hours most weekends, having no children at home, listening to lectures by often very impressive lecturers on every subject under the sun. The most popular lectures were reported to be: "Is there life in outer space?" and "The Bermuda Triangle." It was the questions to the lecturers on issues of importance to which I paid closest attention.

## Afghanistan

The most startling questions concerned the Soviet invasion of Afghanistan. Rumors in a closed society take on exaggerated importance, and there were reports of hundreds of coffins returning to Leningrad. To hear a tough Russian working man ask the lecturer, "What the Hell are we doing there?" was a bit of an eye-opener, as was the answer of more than one lecturer, explaining that their "socialist brothers" in Afghanistan were in a very difficult situation and had requested some fourteen times that the USSR come to their rescue. The lady director of the Dostoevsky Museum appeared genuinely concerned that American countermeasures threatened to "put the bear in a corner," that is, dangerously provoke Russia. She was particularly upset, apparently, because Soviet paratroopers attacking the palace killed a close friend, a doctor treating the Afghan president reportedly for poisoning.

The public Soviet position on Afghanistan was one of confidence. At a farewell dinner given for Nan and me at a restaurant, the head of the Diplomatic Agency assured me that in "three months" the situation would all be quiet in Afghanistan. A slightly murky crystal ball!

I went down to Moscow to discuss with Ambassador Tom Watson what the American reaction should be to the invasion and agreed that we needed to underscore our concern by measures that would hurt Russia's economy and prestige. Boycott of the upcoming Olympic games was one of our proposals, even though we had hoped to have front row seats at the Olympics. But the people who were really hurt by our boycott were not the Russian *nomenklatura* but the tourist guides who ended up translating technical articles instead of practicing their English and absorbing a sense of a freer world outside.

I was thoroughly irritated by the articles in the Western press and even public statements that were based on pseudo history, charging that the real purpose of the Soviet invasion was to fulfill Russia's historic goal of reaching the Persian Gulf. America goes ballistic when an insignificant Cuba or Nicaragua, not even located on the American border, turns communist; yet Russia becomes the old "imperialist bear" when it sees a genuine danger along its 1,700 mile frontier with Afghanistan—a danger not only of loss of face if it abandoned an ally, but also of having a militant Islamic state as a neighbor of its own vulnerable Muslim republics.

## Delegations, Delegations

From a selfish standpoint, we were delighted that the Soviet invasion of Afghanistan brought an abrupt halt to the delegations that kept flooding into Leningrad by train or plane, usually at 8 a.m. on Saturday mornings. But some of the delegations were quite an eye-opener. There were the ferroconcrete specialists praising Russia for its ferroconcrete at a time when all we knew about Soviet concrete was the wire mesh attached to buildings to catch falling cement. Another delegation told us how much better computerized the Leningrad subway system was than our own.

I particularly remember the senatorial delegation of Abraham Ribicoff and Henry Bellmon and the dinner given them by the feisty little ideologue of a first secretary in Leningrad, Grigory Romanov. Romanov rudely interrupted the simultaneous interpretation of Ribicoff's rapid-fire remarks to complain that the interpreter had diminished the graves of the victims of the blockade by speaking of their *grob* (simple graves) rather than *mogila* (tomb): "As an interpreter, you should know the difference." And when Senator Bellmon tried to change the subject diplomatically to trade, Romanov basically said, "We don't need your trade." Later I heard Romanov say to the senators that they should not listen to what their consul general in Leningrad wrote about his city. He said it as though he had been reading my dispatches, which was probably true.

Romanov was one of Mikhail Gorbachev's rivals for the position of general secretary. His insistence on using some of the Hermitage's best dinner service, which was then broken, on the occasion of his daughter's wedding, was typical of this feisty, egotistical Bantam cock of a man. If Romanov had become general secretary, the Cold War could have gone on for another decade. For it was not the economy that destroyed Russia; the Russians had suffered stoically through much worse times. Rather, it was Gorbachev's miscalculation that he could modernize a communist society and make it competitive with capitalism without losing control of the process. Given a choice, probably a majority of Russians would have chosen the Chinese path to modernization—economic liberalization under tight political control.

I cannot say that we were always impressed by the quality of our congressional delegations. One senator sent us three "Confidential" cables to be sure that he got a double bed for his visit with his new bride. Another ignored the advice of my deputy and insisted on changing dollars on the black market. I was a little startled by the purple hair of that old West Virginia senator, Robert Byrd, when he arrived.

But Byrd later did me a favor by instructing me to arrange a concert for one of his constituents over the weekend. She had allegedly sung in some of the great opera houses. I would have liked to know, however, what roles she had sung, and her age. I managed to arrange a concert at the conservatory for some one hundred music students. There I found myself the interpreter for a singer who seemed disturbingly lacking in confidence for someone with her alleged reputation. And then she began to sing. Having had a mother who was a singer, I could appreciate a gorgeous voice. After she had ended her recital with Verdi's *La Forza del Destino*, the bald leading *basso profundo* of the Kirov Opera came over enthusiastically to check our singer's diaphragm. The awed students asked her how old she was, and she said she was 65—still a remarkable voice. I cabled Senator Byrd, "Please send me more such constituents."

## A Rich Cultural Life

The Residence justified itself as a venue for representational events, from classical quartets to the New Orleans Preservation Hall Band, with its 80-year-old leader standing unsteadily on a chair, weaving, and blowing his trumpet. One of our best evenings was showing American ballet on television over a buffet supper: Martha Graham, Twyla Thorp, Alvin Ailey, and others. Valya Ganibalovna, our close friend from the Kirov ballet, brought together some six or seven performers, including Maksimova from the Bolshoi. We went regularly to see Valia, a prima ballerina, dance at the Kirov, becoming one of the competing claques of balletomanes clapping wildly for their own favorites. By chance in 1980 when Nan and I were waiting in St. Petersburg to sail on our Foreign Service friend Alan Logan's ketch in the Baltic, we were presented at the famous Astoria Hotel with a brochure advertising a performance by Valia. She was performing that night with her own company made up of many defectors from the Kirov. After the show, we went to Valia's new apartment, where we dined on vodka and rich cream from their own special merchant, and shared the food with a pet white rat that did little to add to our appetite.

It is hard to describe the emotion and sense of history that we experienced at the seventieth anniversary tribute to Konstantin Sergeyev, one of the grand old men of Russian ballet. Some of his very traditional ballets were performed, and the stage was covered with flowers. We used to watch Sergei and his wife Dudinskaya teaching the future generation of great Russian ballet dancers who had been selected by a rigorous exam to attend the Vaganova School of Ballet in Leningrad. By chance on Channel 98 in Washington in 2008 we saw the two of them dancing in the 1930s, revealing a brilliant, somewhat stocky but lithe contrast to the dumpy figures they had become.

We benefited from the fact that the daughter of the leading cultural official in Leningrad worked at the Consulate. We were given a cultural escort from the Diplomatic Agency, who arranged for us to enjoy a wide swathe of the city's cultural life at the ridiculously low Soviet prices. At the Gorky Theater we saw a spectacular presentation of Chekhov's play *Dachniki*. We were shown the apartment of a man who claimed to have had made his money during the New Economic Policy and used it to buy paintings. He said he had shifted his focus from well-known artists like Repin and Serov to much more unusual paintings and engravings of tsars and members of the nobility—whom some Russians refer to appropriately as the "beautiful people." The paintings covered his walls. On one occasion we escaped from our escort to join a well-known theater critic, who took us to the gravesite near her house where the famous Russian poetess Anna Akhmatova was buried. A close friend of Akhmatova, the critic recited by heart some of her poems over her gravestone.

Our closest Leningrad friend was the sculptor Grisha Israelovich, a member of the Artists' Union and son of a high party official who had run a sanitarium on the Black Sea. Grisha said that he had run a cultural center in Norilsk, the rich mining and gulag center north of the Arctic Circle. Being Jewish, he might well have been one of the inmates. He was a fine sculptor in bronze and introduced us to other sculptors like himself who were testing the limits of political and artistic expression. One of his friends had sculpted a large statue of blindfolded Justice. When it was rejected on ideological grounds, he renamed it Allende, the left-wing leader of Chile whose overthrow was reputedly aided by the CIA—and it was accepted. Since Grisha knew no English and was the center of an active intellectual salon in Leningrad, I advised him at his age not to try to join his divorced wife and daughter in the United States. After we left, his handsome circular stone tower where he did much of his work was taken away from him.

The Dimitrievs were part of this same circle of artistic friends. Viktor was a dissident who published underground *samizdat* publications. The KGB was happy to give Viktor and his half-French wife Mila permission to emigrate. We took them to Disneyland when they arrived in Los Angeles. For a time they taught Russian at Oklahoma, before they separated, and Mila went to work for NASA in Houston, Texas, teaching English to the cosmonauts and Russian to our astronauts. Their daughter Katya married Russia's leading cosmonaut in a space wedding, with the best man an American astronaut wearing a bow tie on the Space Station, and Katya on the ground clutching a *papier-mâché* figure of her groom. They later had a formal wedding in Yaroslavl and live in Moscow.

## Our Soviet Locals

As you might imagine, the Soviets assigned very pretty and smart women to work in the Consulate. One, reputedly the daughter of the Leningrad official responsible for culture, worked in the General Services Office and was so blatant in her effort to seduce the young FSO working with her that his wife put a screen between them. At our first Marine Ball in 1977, she and the man she presented as her husband were the most attractive couple present. To fire her would have caused unnecessary complications with the local government and deprived us of an employee who was very knowledgeable and efficient in caring for visiting delegations.

There was constant pressure by Diplomatic Security to replace our local Soviet staff with contract Americans. But I personally felt it was a mistake. Soviet locals were sources of information about Russian society, a political barometer of our relations, and an opportunity for officers to speak Russian. When we moved to an entirely American staff, we lost both expertise and contacts. And we acquired a false sense of security thinking that we could talk frankly with American contract employees, some of whom were probably in contact with the KGB.

## My Parish: The Baltic States

An important part of my function as consul general was to visit the towns within my consular district that I could get permission to visit. Our competent and pleasant chauffeur, Valentin, would normally drive us down to the capitals of the Baltic states of Riga, Tallinn, and Vilnius, to which I was accredited. I was not authorized to deal with any official above the rank of deputy minister, but I could fly the flag. And the local population reportedly saw this as a sign of our continued interest in their future.

The Russian and Swedish forts facing each other across the Narva River en route to Tallinn were a reminder of Peter the Great's determination to get an ice-free port near Saint Petersburg. Today the Russians are trying to build up their own port near Saint Petersburg so as to reduce their dependence on shipping through the Baltic states—a reflection of their tense relations since those states joined NATO.

A visit to a fishing collective farm in Latvia was memorable for the comment of its director regarding the pollution in the Baltic. I had tasted delicious fish livers in Archangel and asked the director if he sold them. He pointed out that it takes twenty-two years for a drop of water to circulate around the Baltic Sea and that all the pesticides that flow into the Baltic from agriculture settle in the fish livers.

We admired the beautiful sand beaches at Jurmala outside Riga, where the Latvian Central Committee had a dacha. We were told that the dachas in the area were unusually well maintained for Russia because they were privately owned, a phenomenon that we also saw in Tallinn.

The high point of our several visits to Tallinn was certainly the music festival that takes place every five years. In a huge outdoor stadium, a total of 36,000 singers from choirs all over the Baltics sang as one voice. The most impressive song was a nineteenth-century poem to freedom, which had produced riots at earlier festivals. The authorities finally agreed that this nationalist symbol could be part of the program.

Driving back from Tallinn after watching the Olympic sailing trials, Nan was at the wheel around 11 p.m. when she suddenly swerved to avoid a dead cow in the dark. I was less lucky some months later in Leningrad when I was suddenly driven across the intersection at a red light on Nevsky Prospekt by a huge truck, ironically called "automobile repairs." The driver, perched high above me, was too busy looking at this strange foreign car below him to see the red light. Ingostrakh, our Soviet insurance, paid for the repairs in Finland, but it was the Diplomatic Agency in the end that came out ahead, buying my Mustang at a bargain price when I was about to leave Leningrad.

**The Arctic North**

My trip to Archangel was memorable in a very different way. The custom was for the mayor of a city to give a luncheon in my honor, and I would reciprocate the following day. But in this case the mayor was leaving for Moscow the next day so that we held back-to-back banquets, drinking toasts to *Mir I Druzhba* (peace and friendship) until 10 p.m. At that point I was conned by the official accompanying me, who said, "Don't we have something important to discuss?" It turned out that he was a lush who wanted to drink my Black Daniels.

Bleary-eyed, I found myself at 9 a.m. next morning trying to formulate intelligent questions to the director of the local pulp plant. For once I welcomed the ritual glass of Cognac. On my next trip to Archangel I was refused permission even to visit a park of old wooden buildings outside the city, perhaps because I had asked to visit the sensitive naval base at Severnoy Dviny. What I recall most vividly were the crazy angles of the wooden log buildings as their peat foundations sank into the melting permafrost.

Nancy accompanied me on my first trip to Murmansk, providing some protection from the heavy drinking for which the city is known. The most moving moment of our visit was to the small cemetery, perhaps spruced up for us, for the English and American sailors who died in World War II trying to supply Murmansk. (In 2010, illustrating the evolving new Russia, the consul general's residence displayed Leningrad orphans' paintings of ships supplying Murmansk in World War II.).

From Murmansk, we then headed south to a mining area that produced the potash the Soviets exchanged in a deal with Armand Hammer for urea. (Hammer was known in Russia as the man who had shaken Lenin's hand after negotiating a concession for a pencil factory shortly after the revolution.) In Ventspils, Latvia, I later visited the large warehouse that Hammer was building to store the potash before shipping it to Florida. My main reason for going there was to look into complaints from the American workmen regarding their difficulty with the police when they tried to date the local girls. The plant director was quite sympathetic when I explained to him that these virile young workmen were naturally interested in meeting some women. The situation must have improved because later I found myself trying to dissuade one of the workmen from marrying what appeared to be a local prostitute.

## Morale at a Hardship Post

Leningrad was a more difficult hardship post than, say, one in

Africa because of the isolation enforced by both sides. In contrast to the prewar period, Americans were not allowed to fraternize in any serious way with the "locals." And the Soviets set out to control all relationships. There was little to do for the staff that did not read and understand Russian. In Gabon by contrast my secretary learned enough French that she ended up marrying a French businessman. And in Moscow non–Russian speakers at least had access to a large diplomatic corps, with their secretaries and nannies. After some incidents, Marine tours in Leningrad were cut and the Marines sent up to Helsinki every few weeks on the pretext they were to improve their shooting skills, of one kind or another.

Fortunately we had a dacha 45 kilometers outside town, reputedly the "love nest" of Tsar Nichols and a ballerina friend. We encouraged staff members to spend the night there, and picnics were frequent in the summer. At one party in winter, to which we had invited Russian artists, we had the challenge of starting up our cars in the subzero weather. Our indomitable *babushka*, who took care of the dacha, came out carrying a large shovel covered with red-hot coals, which she placed under the car, and we started without blowing up.

## Religious Life

In some countries, lonely staff members who were religious could find an outlet for their emotions, but this too was limited in Leningrad. Besides the Russian-speaking Baptist church to which I earlier referred, there was a basically Polish Catholic church, which my French colleague attended. And about once a month Eric Staples, a wonderful Anglican minister who had been a captain in the Royal Navy and was stationed in Helsinki, would travel through his "tiny" parish, which consisted of Scandinavia, Russia, China, Gibraltar, and one of the queen's churches in London. Our staff was not particularly religious, with the consequence that Nan and I would sometimes find ourselves the sole communicants at the service Eric held at the Residence. I inherited the tradition of a formal service at the Residence; but in retrospect the services might have been better attended if they had been held in the Consulate.

Nan and I found that attending the Russian Orthodox service

was both a religious and an aesthetic experience. At one famous old seminary in Leningrad, great musicians like Borodin and Rimsky-Korsakov were buried. One Easter I was invited after the service in Saint Nicholas Cathedral for an early breakfast in the seminary, presided over by Father Anthony, the new metropolitan of Leningrad. (I was fortunate to have been on leave when the memorial service was held for his controversial predecessor, Father Nikodim. So it fell to my poor deputy Frank Crump and his wife to stand for eight hours in a crowd of invited guests until the service was over.)

The Finnish dean of the consular corps, Ambassador Antti Karpinnen, who had served eleven years in Russia, had advised me that the metropolitan was believed to be a descendent of the noble Naryshkin family, who had taken the name Miller for self-protection. Father Anthony's great love, apparently, was for his ceramics. I took the first opportunity to pay a formal call on Father Anthony and concluded that the rumors were correct. The metropolitan carefully turned around a great ceramic egg to show me the Tsarist double-headed eagle on the back.

I invited him to lunch and to my great surprise he accepted. He arrived in full church regalia, miter and crosier, with the dean of St. Nicholas Cathedral, similarly attired. Our militia guards were bug-eyed. He asked about the history of the Residence (as though he didn't know). When I said that it had belonged to the uncle of the tsar he replied, "Oh, yes, Konstantin Konstantinovich, a great poet," and proceeded with that special Russian facility for remembering poetry, to declaim long passages from the uncle, whom I knew primarily as a writer of religious tracts. If the Residence had not belonged to the tsar's uncle, this would have been a further example of the lengths to which the Soviets would go to perpetuate a self-created myth.

It was probably no coincidence that Father Anthony was a much more conservative metropolitan than his predecessor. When I asked him what he thought of the idea of translating the church service from Slavonic to Russian, as pushed by Father Nikodim, he brushed the idea aside, saying, "Of course, the parishioners could understand Church Slavonic."

## My Hobby

In my spare time I tried to track down where our ministers and our few ambassadors to Saint Petersburg had lived in the city. My interest had been piqued by my search for consular property, and my discovery that the U.S. government had never owned but always rented property—or to be more exact, each envoy had rented according to his pocketbook. Our first minister, John Quincy Adams, who at a young age had accompanied Francis Dana on his abortive mission to Catherine the Great, had first lived in a lovely house on the Neva, next to the summer garden, and then across the square from Saint Isaac's Cathedral. He soon found that on the miserly salary that Congress authorized its diplomats, it was impossible to keep up with the other members of the diplomatic community, many of noble families, and was finally forced to move "below the salt," to the disgust of his wife. Tsar Alexander I, who apparently liked Adams, is reported to have commiserated with him regarding his financial state.

A wealthy southern slave owner, Henry Middleton, was selected to represent the United States in the dispute with the British over the seizure of American slaves during the War of 1812, which Alexander I was chosen to arbitrate. Given Russia's serf economy, and perhaps recognizing that America was a growing naval power that might help reduce dependence on the British, Alexander ruled in favor of the Americans. As one might infer from the beautiful Henry Middleton estate in Charleston, South Carolina, Middleton had no financial problems during his ten-year tenure as minister, (1820–30); but he lost two children to the pestilential Saint Petersburg climate.

General Cassius Marcellus Clay of Kentucky (1863–69) was another wealthy American sent as minister to get him out of Kentucky, but he enjoyed his stay so much that he came back for a second tour. A Russian noblewoman is reported to have descended on Clay in Kentucky on New Year's Eve and handed him a bundle telling him, "Cassius, here is your son."

What used to be in my day the beautiful Wedding Palace, with its Carrera marble staircase, along Krasnopressnaya (now under its old name, Angliskaya, or English) embankment, was reportedly decorated by its wealthy renter, the American ambassador, Chicago businessman Robert McCormick (1903–5). I could never find the house of James Buchanan (1832–33) along the Neva River. Buchanan had been quite effective as minister to Russia, but was a weak president. The trade agreement Buchanan negotiated while minister to Russia lasted until 1913, undone by a familiar issue, the treatment of Russian Jews by the tsar.

Initially when I went to the Saltykov Schedrin Library to do my research, the librarians were very helpful. But the next time I asked for the same old maps and publications, it took them forever to produce them, presumably convinced that there must be some secret messages being passed through this curious pretext of research. And one evening, with the temperature perhaps 30 degrees Fahrenheit below zero, I emerged to find that my expensive mink *shapka* was not at the *garde robe*.

We had faced quite a bit of harassment, and I was convinced that this was a ploy to get me to give up my research. I accordingly protested this incident loudly and persistently to the Diplomatic Agency, which took up the matter with Moscow. After some three months, I was presented with what I call my KGB mink *shapka*, of better quality than the one I had lost. Meanwhile, I had returned to the library with a different hat of sheepskin and warned the elderly guardian that I did not want a repetition of what had happened before. Indignant he said, "Where did you leave your hat?" and I pointed at the *garde robe* at the other side of the room. "Of course," he sneered, "those young ones probably sold it for a bottle of vodka." I suspect he was right. But I still wear my KGB *shapka* without a guilty conscience.

### A Feisty Scot

Foreign Service life can be enriching or disappointing, depending not so much on where you are assigned as on the character of your colleagues, and particularly your ambassador. I was very fortunate in getting along well with my ambassador, Malcolm Toon, from his first day as political counselor in Moscow in the 1960s. Born in Edinburgh, he was not everyone's cup of tea. He had little patience for fools or complainers and expressed his strong opinions with blunt, undiplomatic frankness.

Toon related how he had asked Henry Kissinger why he would be sending him to Israel, of all places, to which Henry had replied, according to Toon, that he had "looked for the biggest SOB in the Foreign Service to deal with the Israelis and do to them what they do to us in their blatant lobbying of the Congress." Warned by Ambassador Kohler that I might be a problem, Toon said he had told Kohler that "he could handle any Scot." More important perhaps, he and his wife Betty became very fond of my beautiful Nancy.

We had talked often when Toon was counselor of taking a trip to Siberia together. But the opportunity did not come until Toon became ambassador in Moscow and I was consul general. It was around 1980 when the Toons and Buchanans flew out to Khabarovsk on the Amur River. We had a great time discussing preservation of the forests and the Siberian tiger with an unusually frank party secretary, watching a revealing fashion show with native décor, and walking around the fishermen sitting drinking over their holes in the ice on the Amur River. On the Trans Siberian Railroad we rolled through low hills and past frozen rivers used as roads for truck traffic in winter. In legendary Chita, the brown coal used for heating quickly drove us back onto the train. Some Aussies commented to me: "We notice you chaps are eating mighty well." And so we were. The chef asked each morning what the ambassador would like to eat that day. We were treated to translucent red caviar the size of small grapes.

Lake Baikal in winter, with its pure water frozen to a depth of two meters, as though carved by a sculptor with snowflakes imbedded in the ice, was the high point of our trip. Baikal contains a quarter of all the earth's fresh water and is among the deepest lakes in the world. Like the rivers, Baikal was a major truck route in winter. Soviet scientists and environmentalists succeeded in persuading the Kremlin to remove the pulp plant that had been polluting the water, thereby saving its unique forms of fish (such as the Omul) and freshwater seals. President Dmitry Medvedev apparently did not approve of the decision of his prime minister, Vladimir Putin, to authorize the mill to resume operations.

I could measure the environmental progress by an earlier visit I had made to the mammoth Bratsk dam on the Angara River, which flows out of Baikal. Having heard of the terrible mosquitoes that made working on the dam almost unbearable, I asked what they did about them. It was very easy, I was told. Since the mosquitoes hatched behind the boulders in the Angara, the Soviets dumped barrels of DDT pesticide into the water—with no concern apparently for the environment.

At a little wooden church on the shore of Lake Baikal, we met a young man leaving to serve in the army who hoped to return one day as a priest to this church. It is in the beautiful cathedral of Irkutsk near Baikal that Princess Trubetskoi is buried. She had followed her husband, the incompetent leader of the December 1825 comic-opera revolt, into exile. When our dear friend, Olga Davydov, a direct descendant of the Trubetskoi family, was asked to speak about her family at a meeting in Russia of the Decembrist Society, she found that most of its members knew more about her family than she did. She was also related to the progressive Tsarist Minister Davydov, whose statue stands outside Novodevichiy monastery in Moscow. When she visited the statue and talked to the *babushky* outside the church, they chided her for not cleaning off the statue of her relative.

Back in Leningrad, I was informed that, by unanimous vote in Moscow, I had been named master of ceremonies for a going away party for Ambassador Toon. As much as I would have liked to decline the honor, I made an effort. Dressed up like a Soviet citizen in a blue suit, medals, and wig, I read a poem, "To a Toon from a Goon," imagining what Toon's KGB "goon" might have written about the ambassador in his dossier.

# 14

# Retirement?

Toon was called back from retirement around 1992 to head an effort to locate American MIAs who may have died in Russia or might still be alive. This was an effort to satisfy Senator Robert C. Smith from New Hampshire, with little confidence on the part of Soviet experts that there would be any useful result. Toon still gave it his best effort and found a surprisingly helpful counterpart in General Dmitri Volkogonov, a critic of Stalinist Russia.

By agreement with the Russian government, Toon would address the population in different cities in Russian, asking them for information they might have about Americans living or imprisoned in Russia. He suspected without firm evidence that the KGB had taken fliers downed in Korea back to the Soviet Union for interrogation and eventual execution.

I asked about the case I remembered of an American reconnaissance plane that was shot down off Vladivostok, with the crew seen being pulled out of the water. Here he had unexpected success. A Russian appeared hoping to sell a Navy Annapolis ring belonging to one of the officers of that ill-fated plane. He said that he had pulled the American flier out of the water, but he was already dead; he buried him on a neighboring island and took his ring. Eventually the body was found and returned to the family, who invited the Russian and his family to attend the funeral service.

My own retirement was not long in coming. After Leningrad in 1980, I was assigned to the Army Defense University, where I worked on issues involving the Soviets in Africa. I no longer had good contacts in the Africa Bureau; and when I asked whether there was any chance of a post in Africa, it became plain that I had burned my bridges. I was told curtly, "We understood that you did not want to serve in Africa." *Sic transit* Buchanan's career in the Foreign Service. I was, in fact, ready to leave a Foreign Service that seemed to me to have lost some of its *esprit de corps*. If I had been more patient and persevered in the Service, I might have qualified for one of the posts that opened up in the former Soviet empire following the collapse of Communism.

**Foreign Service in Perspective**

This seems like an appropriate moment to look back on my years in the Foreign Service, weighing it on the scales of life's values. I must say at the outset that I have had a generally happy and lucky career. I agree with my wife that we were there in *la belle époque*, by which we mean a time when the old traditions of the Foreign Service had lost their sometime stifling pomposity but still provided a reassuring and often gracious framework for diplomatic life. I am doubtless prejudiced, but I like to believe that the *esprit de corps* of the Foreign Service in my day was greater than it is today, when it must recruit for a much broader range of specialties to deal with a much more complex world than in my day. I was also unusually fortunate in being assigned either to posts on my wish list or to pleasant surprises like Paris and Norway. And with few exceptions my ambassadors and colleagues were experienced, competent, often amusing and decent human beings, good companions on a sometimes-difficult journey.

Though I entered the Foreign Service "laterally" via the Wriston Program, I was never made to feel a second-class citizen. I believe that the writing and research discipline of my early years in the Intelligence Bureau compensated for the years I missed as a junior FSO rotated to get some feel for work in all sections of an embassy. But there were times when I wished that I had had some consular or administrative experience.

My one serious consular challenge came in Burundi when a huge, self-proclaimed liberation fighter against Portuguese colonialism in Mozambique asked for a visa to visit the United States His passport showed that other U.S. embassies had granted him a visa, but there was something about the man that disturbed me. So

I cabled Washington asking if they knew anything about him, and the word came back that he was not an African but born in Texas. Yet he spoke African English and, I believe, had convinced himself that he was an African liberation fighter.

Though much will have remained substantively the same in a modernized, expanded Foreign Service, what may have been lost on the part of FSOs are a certain *esprit de corps* and a feeling for and understanding of the countries to which they are assigned:

- You cannot escape the tedium of repetition, writing the same reports, checking the same sources, meeting the same contacts. An important difference is that in my day I had a secretary to decipher my undecipherable handwriting, but now each FSO must be his own secretary using his own computer.

- From the standpoint of creature comforts, life at home and in the office is easier. No longer do families have to deal with a lack of air conditioning or window screens. The new FSOs are almost guaranteed the comforts of home, in a sanitary box, insulated from the outside world. This ghetto existence undermines the very *raison d'être* of being a Foreign Service officer, who is expected to explore the society and people around him. Too much has changed for us to go back to the time when missions were not top heavy with administrative personnel and there were many more small consulates staffed by officers living on and in the economy, and therefore much better informed about what was happening in their country of assignment than their counterparts today. I suspect most FSOs, particularly those with Peace Corps experience, would be pleased to run the risk of living in the community rather than being condemned to living a Green Zone–type existence. There are certainly places where living out in the community would be foolhardy. But they are the exceptions. Our housing is designed on the false assumption that they are the rule. The bureaucrats responsible for security and building our embassies have acquired too much influence.

- You learn fairly early to bite your lip as your superiors take credit for your ideas, or not to question what seem like unreasonable assignments. It was flattering, for example, to be asked to complete a complicated assignment overnight—in my case, Berlin—or over a weekend for Foster Dulles on situations of confrontation with the Soviets. But you cannot help but grumble that, with a bit of forethought, the product could have been much less onerous to produce and probably better. Employees of all professions face similar pressures.

- One of the most challenging aspects of Foreign Service life was juggling one's professional responsibilities with those toward one's spouse and children. In couples I have seen, the wife saw herself as a simple appendage of her husband's career, living a life she increasingly found distasteful. In the close circumstances of Foreign Service life, the unhappiness of one partner inevitably affects the performance of the other. I can imagine in the new Foreign Service, where the accompanying spouse may be the man, it is particularly important to find some meaningful activity for the spouse, as in the example I cited from Burundi, where the spouse kept the motor pool running. But even successful Foreign Service families, where both partners entered the service with a spirit of excitement and acceptance of whatever hardships life might have in store for them, have found it difficult to manage the competing priorities of work and family.

  Given representational responsibilities in the Foreign Service, juggling one's work responsibilities with those toward one's spouse and children may be more challenging than in many other professions. Our own children appreciated the advantages of life abroad but resented, we later learned, what seemed to them to be the priority we gave to the Foreign Service. Each Foreign Service family is unique, and it is encouraging to see how many Foreign Service children go on to have a successful career as Foreign Service officers.

- Certainly the most unpleasant chore in Foreign Service life is writing the annual OER (Officer Evaluation Report), particularly when the person about whom one is writing is an experienced, competent, but not brilliant friend. There is a natural tendency to gild the lily in these reports and omit any reference to weakness, which could help improve performance but which in an OER can be a death sentence with the Promotion Boards. This understandable inflation of OERs by the rating officer is hard to combat when any superior knows that to get a deserving employee promoted he must compete with reports even less honest than his own.

  The only time I served on a promotion panel for Class III officers, we were required to rate from 1 to 300 the three hundred officers whose dossiers we were given. Inevitably we looked for any small flaw that might distinguish one paragon from another. The lowest 10 percent were subject to being "selected out," and the top 10 to 15 percent might be promoted, if there was any money for promotions. One case involved a brilliant but (reading between the lines) obnoxious officer, the type to disrupt any small post. Half of the panel wanted him selected out, but the other half felt he deserved promotion. My cousin, who became a brilliant CIA analyst on Southeast Asia, was selected out because an old curmudgeon FSO in Cambodia destroyed his career on his first assignment.

- Given the competitive environment in the Foreign Service, with officers struggling to get ahead and avoid being selected out, it was surprising to me that the *esprit de corps* was as good as it was. The insecurity of competition often brings out the worst in otherwise decent people. A superb officer who worked for me in Moscow would probably have been named ambassador to Poland where he had served were it not for a rival who spread the lie that he was "soft on communism." Of course, these things happen in every profession.

Would I advise a young man or woman to enter the Foreign Service? It would depend very much on whether the individual was motivated by a spirit of public service and adventure and a curiosity about other cultures. The Foreign Service is not for those looking for financial reward or glamour. The prospective officer should preferably be the type of person who is more amused than irritated by the frustrations of bureaucratic life or of life in a Third World post, is courageous without being foolhardy, and has good judgment under pressure and the personality and motivation to make friends and influence what are often young leaders of their own age in the Third World.

While many FSOs are like me and prefer service overseas to the bureaucratic rat race of Washington, careers are more often made in Washington than in the field. A Washington assignment is important, not only because of the contacts you make, but also because you learn how to operate in a ruthless, competitive interagency environment and to appreciate the importance of domestic politics when making any foreign policy decision—considerations of growing importance as you move up the diplomatic ladder.

I could only wish for a new Foreign Service officer that he or she would have as engrossing, fulfilling, and enjoyable a life as I have had, with whatever the traumas in professional or personal life, which add to wisdom and memory.

# PHOTO GALLERY 3

Tom in retirement

Sailing in the Mediterranean with retired FSO Alan Logan, his wife Nicole, and Nancy

Skiing in Dombai in the Russian Caucasus, 1984

Repairing helicopter that was to drop off skiers in Dombai

At 61 on top of the 17,000 foot Umashila Glacier in the Indian Himalayas with monk porters, 1985

Nurses at hospital in Dombai thank us for delivery of medical supplies from Jackson, Wyoming, 1984

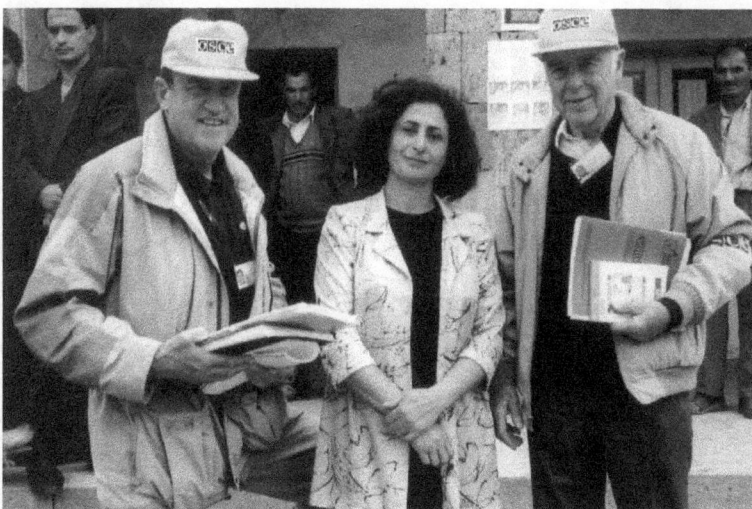

As OSCE inspector in Armenia in 2000

Village of Saumane de Vaucluse,
Provence, France

Nancy and Tom's house in Saumane de Vaucluse

Loggia in their house in Saumane de Vaucluse

Author's sister, Babs, daughter Barbara, and son Campbell in Wyoming, 1995

Tom and Nancy before their cabin in Jackson Hole, Wyoming, 1995

View of the Potomac River from Nancy and Tom's patio in Maryland, where they lived from 1951 to 2000 when home

# 15

# Back to Russia

In 1981, the year of my retirement, I received a letter in France asking me to do a study for INR on Soviet "active measures" (political warfare) in Africa. Then I was asked to analyze Soviet diplomatic assignments to Africa as to who were real diplomats, who KGB, and who disgraced politicians. Finally I joined an army of retired FSOs working on Freedom of Information, helping to declassify classified documents as required by congressional and executive legislation—or, one could say, declassifying the documents one had overclassified years ago.

## Processing Refugees

In 1990 I spent several months back in Moscow working this time for the Justice Department, or more precisely the Immigration and Naturalization Service (INS), processing would-be refugees. For certain categories of applicants, refugee status was almost automatic. It was presumed that Soviet Jews, Pentecostals, and members of the Ukrainian Auto Cephalic Church had suffered persecution. Our job was to determine that persecution had taken place and that the applicants were who they said they were. It would have been better to have had consular rather than interrogation experience for what became a sort of visa-mill job.

The Jewish applicants came suspiciously well briefed on what to say. It was almost amusing to see Russians trying to convince us that they had a Jewish grandmother when for decades they would have hidden the fact, if it were true. The Pentecostals, usually less educated but franker, inspired more confidence. But there was one case involving would-be Pentecostals that I still wish I could redo. An attractive young couple with a lovely daughter, all speaking good English, claimed to be Pentecostals, having heard that this was the key to refugee status. They told me when they had their water baptism, but when I asked about "spiritual baptism," when the Holy Spirit comes down and you speak in foreign tongues, they looked helplessly to Heaven. They were clearly not Pentecostals, but they would have contributed so much more as citizens than all those old folks going to join relatives in the United States.

## Ukrainian Nationalism

We took a trip to Lvov (now Lviv) in the Ukraine, where we visited a relative of one of our many Ukrainian reviewing officers. It was an exciting moment. The 94-year-old Patriarch Mstyslav of the Auto Cephalic Church had returned from the United States and held a religious service that overflowed the church out onto the square. Russian Orthodox priests complained hypocritically of the behavior of the Greek Catholics and Autocephalists in taking back the churches that had been illegally taken over by the Russian Orthodox Church. The monument to Lenin had been torn down, and as we stood bareheaded in the freezing cold, all of Lvov seemed to be singing the Ukrainian national anthem. "We will show those people in Kiev what it means to be a real Ukrainian" was a common refrain. I had never felt such nationalist euphoria.

## Sleeping with Stalin

When Nancy, or Nanya as I sometimes call her, Russian style, joined me in September 1990, an embassy local suggested that we might like to spend a week at Gagra at what was formerly Stalin's dacha. Gagra, near the Georgian separatist enclave of Abkhazia, under Russian protection, was also Khrushchev's favorite resort on the Black Sea. We arrived at night in the pouring rain in Sochi, the future site of the Winter Olympics, and after a 45-minute taxi ride discovered we were not expected until the following day. But we lucked out and were accepted at a sanatorium, obviously meant for the *nomenklatura*, giving us an insight into a different lifestyle, but one which clearly did not include a place to hang or store one's clothes. Stalin's dacha was quite modest, with a housekeeper who fondly remembered Stalin's concern for his fruit trees. We supposedly slept in Stalin's bed, without nightmares.

## Abortive Putsch

In 1991, at an historic moment, I worked briefly for INS. Early in the morning of August 21 I was awakened in my room on the twenty-first story of the Ukraina Hotel by the sound of tanks coming down Kutuzovsky Prospekt. Happily, the tanks were coming to support the government holed up in the White House, the Parliament Building across the Moscow River from my hotel. I did not see Yeltsin speaking from a tank, but I walked through the barriers set up on the bridge to protect the White House and saw the flowers placed by the underpass near the Chancery on Tchaikovsky Prospect to mark the spot where a tank had inadvertently crushed three students. Political activists and young democrats came out to protect the government, but the bulk of the Moscow population waited, Russian style, with their fingers in the air, to see which way the power struggle would go.

## Aid to Russia

In 1992 Ambassador Robert Barry, an old Soviet hand, asked me to go to Moscow for a couple of months to help distribute aid in the form of supplies of all sorts that had been stored in military depots in Western Europe. The Russian economy had collapsed, in part because of unrealistic economic advice, particularly from U.S. economists trying to implement a crash program of economic reform like that which had succeeded in Poland. There was plenty of blame to go around, but the reality was that little old ladies, some obviously from the middle class, were standing on street corners trying to sell a bottle of vodka or a package of Marlboros to survive. The Russians claim that their depression was worse than the Great Depression of 1929. One needs only to have lived through this period to understand the enormous popularity of Prime Minister Putin, who brought stability out of chaos so that salaries and pensions were again paid on time.

The supplies being flown in from Western Europe were more of a symbolic gesture than meaningful aid, but they were welcomed. I worked out of one of the former Gorbachev press centers and spent much of my time standing in freezing weather at the airport supervising the distribution of supplies to agreed recipients. Even our foul-ups were appreciated.

A circus group from Rostov was most persuasive, but when I insisted that someone must accompany the shipment to Rostov it turned out that it was Rostov on Don, thousands of kilometers away, and not the monastery town of Rostov Veliki near Moscow. The circus toughs persuaded me of their real need and that they would deal with anyone trying to steal their shipment on the train.

I accompanied one shipment to a prison, reported in dire straits. We arrived at night after a four-hour drive. The young male prisoners were very efficient in unloading our truck. It was only then that I could read the labels on the boxes. To my horror they were marked as items for women. The very nice commandant of the prison said, "Don't worry, I have 250 women on my staff, and they will be very grateful." And knowing that prison guards in Russia often did not live much better than their prisoners, I had no compunction about handing over our mislabeled shipment.

I was asked to stay on for a few months to work with USAID on humanitarian aid. The work took me away from Moscow on two assignments. Montclair, New Jersey, asked us to check whether the population of its sister city, Cherepovets, had enough food. I spent three days in Cherepovets and determined that although there was enough food, what was really lacking were medical supplies of every kind. I talked with the impressive surgeon who ran a thousand-bed hospital and asked why it took so long for patients to get out of a Soviet hospital. He replied that all he had in the way of equipment was one x-ray machine. He had no ultrasound or other modern medical equipment. "It takes me this long," he said, "to figure what is wrong with my patients." I urged Montclair to send medical equipment of all sorts.

One of Vice President Cheney's daughters was working with Richard Armitage in support of USAID. They had decided to send two American executives to try to convert the military research center of Semi-Palatinsk in Kazakhstan, the Russian equivalent of our

Nevada Testing Grounds, into a market economy. It was not clear to me how two American businessmen who may speak little or no Russian and have never worked in Russia were going to achieve this miracle. But mine was not to reason why. My first reaction was that Semi-Palatinsk resembled the old Cheyenne, Wyoming, in its desolateness and the freezing wind mixed with snow for which I was not prepared so far south. I was met by the mayor of the city, a very sophisticated Kazakh who had just returned from a year of studying in Tokyo, having failed to get into the Harvard Business School. Together, we visited a Russian factory that produced amphibious vehicles. The plant was typically rundown and dirty, and the young Russian director treated with clear racial contempt suggestions from the Kazakh mayor about producing civilian vehicles.

From there we visited a Kazakh marble factory. The difference was striking. The factory was clean, neat as a pin, and obviously well managed. I was embarrassed to learn, however, that an American company had not delivered the marble-cutting equipment it had promised. When I inquired briefly in Washington, I was not impressed with the company's explanation.

I was not invited into the military Polygon, with its ultramodern high-temperature research facilities, but I learned that high-tech American firms were already sending over their specialists. The mayor and I agreed that what the town needed were some small industries—brick-making, tomato canning, glass production, and the like—to make the town independent of the large, Soviet-era regional facilities and to provide needed employment. I enjoyed the mayor's warm hospitality, with the exception of the two helpings of tough horseflesh, a local delicacy I felt obliged to eat.

Having developed a certain amount of enthusiasm for the project, even among the suspicious Russian military, I was anxious to know what happened. Eventually I was told that it had been decided that the radiation level in the Semi-Palatinsk region was too high to risk the health of our executives. It was nice to know.

## Cruising on the Volga

In 2006, we cruised down the Volga from Moscow to Volgograd (Stalingrad to me), on a trip organized by an FSO colleague from early Moscow days. And again, in 2010, we sailed the Volga and its tributary rivers, reservoirs, and canals, from Moscow to Saint Petersburg, at the request of our son, who wanted to see Russia again. One used to say that leaving Leningrad/Saint Petersburg or Moscow was like falling off a cliff, so great was the difference in the standard of living outside the two great centers. But the towns that we visited—Samara, Kazan, Ulyanovsk, Nizhny Novgorod, and to the north, Uglich and Yaroslavl—all showed that some prosperity was trickling down from the center, with virtually all the cities having their McDonalds, their Baskin-Robbins, well-stocked gastronomy (grocery) stores with plenty of clients, and the ladies were well-dressed in their high heels and skin-tight jeans. Behind this façade of prosperity, however, one found on the side streets many of the same rundown buildings and trash that we associated with the old Russia. A taxi driver in Saint Petersburg, one of a number, incidentally, driving Fords or Chevrolets, pointed out that most of the handsome old buildings in the center of town still had communal apartments, like the one that he had lived in until recently. On the other hand, all along the Volga we saw a variety of huge mansions built as summer dachas by the newly rich.

Discussing then-President Putin with the fifteen-odd Russian tourists on our boat to Stalingrad, I was not surprised to hear them all say they hoped that Putin would ignore the Constitution and run for a third term in office. They credited Putin with having created the stability and unusual prosperity that Russia was then enjoying. A taxi driver in Saint Petersburg seemed quite content with the tandem leadership of Medvedev and Putin. While Medvedev struck him as being a bit young, he said he would not care if the next election in 2012 saw the tandem unchanged, or reversed, with Putin becoming president again. This appeared to be a prevalent view of the Russian government.

There were clear class distinctions among our Russian passengers sailing south on the Volga. Some had had to borrow against their pensions to be able to afford the tour. But a Russian doctor said casually that she shopped three times a year in Dubai, obviously irritating her fellow Russians. A Russian-American lady from San Francisco, invited for tea by the doctor, returned agog over the opulence she had seen, including the chauffeur, chef, housemaid, and third small low-key car to take their child to school. She said that she asked how many of the doctor's friends lived like her, and she replied with the surprise of some inhabitant of a gated community in America, "Why all of our friends, of course."

At the other end of the economic spectrum, our three attractive interpreters, all graduates of the language school in Nizhny Novgorod, said that they had all hoped to teach but could not live on a teacher's salary. And our gentlemanly porter at Sheremetyevo Airport said that he had been a nuclear engineer, but his company, working on space satellites, had gone belly-up.

I left Russia for probably the last time pleased to see the changes for the better in the life of most Russians, at least those in larger cities. With the end of the Cold War, the character of the consulate in Saint Petersburg had changed radically. It became an unclassified post without Marine guards. It was no longer the besieged outpost of my day, trying to learn more about a hostile empire in the face of official pressure and harassment. The challenges facing the consulate and the opportunities for initiative are totally different than in my day. Saint Petersburg seemed like any normal consulate in a large, attractive urban center with a substantial American and foreign business community, a flood of tourists, and good working relations with local government officials. I envied the opportunity for outreach that our consulate enjoys today. But, at the risk of sounding a bit puerile, there was an excitement we felt when we were on the front lines of the Cold War, which I missed in today's Saint Petersburg.

# 16

# There Is Life after Retirement

Because of the kind of life they had lived, FSOs may be better prepared to take advantage of their retirement years than many Americans with a narrower professional experience. Many will not have
lost their appetite for seeing and learning about unfamiliar parts
of the world. They will have a wide network of friends, foreign
and American, developed at their various diplomatic posts. This
certainly was true of our family. But we were more privileged than
most retired FSOs in having a house in Provence, in a sense an emo-
tionally satisfying extension of our Foreign Service life, a real home
abroad.

## Return to Saumane

We still visited briefly in Jackson Hole, excited by the beauty of the
Tetons, refreshed by the crisp mountain air, visiting old friends.
But Provence was a different dimension of pleasure, more exotic,
sensual, intoxicating to the senses, with the smells of rosemary,
lavender, and thyme mingled in the Sunday market with bright
flowers, the smell of spicy sausage from the Ardeche, the delicate
bouquets of herbal soaps, and of the herbs, basil, fennel, and pars-
ley, and vegetables sold by the farmers. Most important were the
olives. It was a ritual for us every Sunday to visit the stand where
an attractive young couple sold thirty-six different varieties of olives flavored with garlic, lemon, *herbes de provence*, olives of every
description—green, black, the small *Niçoises*, as well as virgin olive
oil. Nuts and herring, too, were sold.

The young olive seller had a *license* (like an MA) in business, and his wife had passed the demanding tests to become a European guide. But they found running his father's olive business, reading and sitting, stirring their huge jars of olives over the winter, perhaps less remunerative but more satisfying. They sold their olives all over the region.

The merchants were as varied and colorful as their wares: black North Africans selling leather belts and bags, peasant women in regional dress, men playing old tunes on a hurdy-gurdy or singing popular songs, and the town crier, accompanied by an ancient terrier, dressed in an old- fashioned costume and a *tricorne* (three-cornered hat), tapping on his drum while he made unintelligible announcements. And located all along the encircling Sorgue River, next to what used to be beautiful *platanes* (plane trees) before they were killed by a blight, were *brocantes* and antique stores of every kind and quality, with the flea market often more expensive than the antique stores.

On weekdays, there was the farmer's market at Velleron, where buyers would rush in as soon as the gate was opened at 6 p.m. to compare *prix et qualité* at stands selling everything from fruits to the wonderful goat cheeses of the region. Since they sold mostly by *plateaux* (trays), it was a good idea to share an expedition with a friend. Afterwards, we would return to our little loggia in Saumane, open a bottle of wine, slice some sausage, and pitch the olive pits down to the garden below, hoping they might take root as olive trees.

### Restoring Our Saumane House

When we sold our original house, we kept the rooftop garden and large grotto, essentially an extension of the garden of our new house. Like everything else in these old villages, who owns what is determined as much by oral tradition as by official survey. So it was that we exchanged the chicken house on the third floor of our ruin for a *voute* (vaulted room) next to our one-lane street, which we later converted into one of the more challenging garages for parking a car.

In the house itself we added a marble bathroom to our high-beamed barn bedroom that led out to a small rear garden with its own little grotto. In trying to cut a door from our kitchen into a lower bedroom that had a cliff for a wall, I found myself trying to cut through a cement supporting wall with a pneumatic drill. Working late one night, I heard an anguished cry from a neighbor: "*Alors les Américains, ça suffit, non!*" (loosely translated: "Cut it out!").

We haunted the *brocantes* for simple furniture to fit our Provençal décor. We learned about French gas stoves, refrigerators, and light fixtures, not to mention plumbing supplies of all types, and pretended to understand our plumber as he explained the functioning of our Rube Goldberg heating unit located in another large vaulted cellar under our garden terrace, next to a large fig tree whose roots extended deeply into the cellar.

## Creating a Terraced Garden

The real challenge was the "garden," or rather the hillside of houses that had disintegrated since the Middle Ages to form a layer of soil on top of collapsed rooms. My vision was to reproduce the terraces dating from the Middle Ages that covered the neighboring hills, formerly planted with olive and almond trees and, in the later years, grapes, when the population was obviously much larger. I had all the necessary stones. What I needed was a good mason, but to find one late in the season was a challenge.

I presented my problem to a new friend, Alain Desmarez (he and his wife made beautiful instant-antique furniture), and he produced Jean Paul, a six-foot-eight, black-haired, beanpole of a Corsican, who had spent eight years in that toughest of all military institutions, the French Foreign Legion. Alain said: "If he likes you, he has a wonderful feel for stone." Happily he drank very little alcohol, and Nan won his heart by feeding him large quantities of tomato juice. He had tried his hand at running an aquarium in Marseilles and was now raising vegetables, selling them at 5 a.m. in the market. He would then appear below my bedroom window at 7 a.m., shouting, "Tom, Tom."

For weeks we worked in the blazing sun from early morning until 7 p.m. Jean Paul was a perfectionist, selecting special stones from our hillside, which he artistically placed to form a double row of dry stone walls, between which I would pour concrete to consolidate the walls. After washing off the concrete oozing between the cracks, the walls looked like dry stone walls dating back to the Middle Ages. It was backbreaking work, because after the concrete, sand, and gravel were deposited outside our garage, they had to be carried up flights of stairs into our garden, where we had dragged a cement mixer. When the time came, we celebrated by planting our terraces with lavender, laurel, and rosemary, which contrasted with the white flowers of our almond trees and the red fruit and flowers of the pomegranate bushes.

Our brilliant idea to dig down and make a wine cellar out of one of the rooms of an old collapsed house proved rather disastrous. When the rains came, the walls of the future cellar collapsed, and I could never insulate it properly, so that the water oozed through the ceiling, disintegrating the labels on my wine bottles, making the serving of wine at dinner a game of probability.

The Russian connection always followed me. Jean Paul hired a strapping young man called Nicholas Pivovarshik. Of Russian origin, Nicholas, who loved beer, did not know that his name meant "brewer of beer." He eventually split with Jean Paul when he found him, as I did, not responsible, nor indeed honest, when it came to money.

Like many Corsicans, I imagine, Jean Paul enjoyed telling stories, with his Provençal accent and emotion. He complained, he said, to the teacher of his son that the boy was not hearing anything about the heroes of France. Everyone was someone called Ali or Hassan. Whereupon the teacher called him a "racist," and he told her, "Yes, madam, I am."

Another incident went back to his time in Vietnam, when he had a small squad of legionnaires, almost all of whom were former German SS troops. On one occasion, conducting review in his Jeep, he jumped up, saluted Nazi fashion, and shouted "Heil Hitler," whereupon to a man the Germans shouted back, "Sieg Heil," saluted, and stopped, shamefaced.

## Our Difficult Neighbors

I took pleasure in telling our French friends who had made fun of our original house with its *pissoir* that we had sold it at a profit to a

Frenchman. I neglected to add that, if we counted the 24,000 francs we paid M. Favre for detailed plans that we never used, then we actually sold at a loss.

The ruin was eventually bought by a M. Gros and his German wife. I thought that I had established a good relationship with Gros over a glass of *Pastis*, and it was agreed that I could raise the height of the wall separating our two gardens. We had looked directly down into what was now their garden, with its mulberry tree dating from the time that silkworms had been raised in our grotto. But Mme. Gros, who turned out to be the real owner of their house, objected to the higher wall I had built, with what I thought was the approval of M. Gros. Accordingly, in October 1982 I received a notice from *l'huissier de justice* (bailiff) of Saumane ordering me to take down the wall.

In the same notice, the bailiff ordered me to remove some construction debris that we had left on a rocky path between my grotto and Gros's closed rear entrance. It was then I learned that the path had been a *voie communale* (right-of-way) but now was a "path to nowhere" once the property was sold. The owner of the lot, Grandvillmain, had neglected to tell me about the right-of-way when he sold me his property, and both we and our notary public had failed to see the tiny thread running through the property on the minuscule *cadastre* (village survey map). If I had foreseen the problems this right-of-way was to cause us, abutting as it did Gros's house and our grotto, where we liked to entertain guests, I would not have sold that end of the Gros house.

Once I knew about the right-of-way, I discussed with the mayor the possibility of buying the right-of-way, which now went really nowhere, and he seemed in agreement. He also agreed that I could put a door between the stone walls at the entrance to the right-of-way to deter children from playing among the dangerous debris of my garden and, incidentally, deter curious tourists from watching us eat in our grotto. Gros promptly challenged my right to have built the door, even though he was given a key.

Meanwhile, Gros showed unusual interest in the end of my garden next to the village square. He had told me that he was going to force the Mairie to move the *pissoir* so that he could extend his garden, and he evidently hoped to own what I thought was my property, the end of my garden. Gros informed me with pleasure that he had had his *géomètre* (surveyer) check the city plat and found that the tiny end of my garden, about 12 feet wide above the city shed, was in fact communal property and that I could not accordingly build the wall that I had planned next to the shed. Again, Grandvillmain, the ostensible owner of this bit of land, had misled me. I thanked Gros and redrew the sketch of my wall 10 feet away from the shed so that it would not impinge on communal property, and I stretched a string across the end of the garden to show Jean Paul where the wall should go, and left for Washington.

In the fall of 1982 I heard from Jean Paul that Gros had complained to the Mairie that I had built a wall 40 centimeters on communal land. It would seem that Jean Paul had built my wall on the wrong side of my string so that the wall was indeed 40 centimeters on communal land. I apologized to the mayor saying it was an understandable error since I was not in France and pointed out that I had also built, at my expense, a retaining wall around a pine tree I had planted on what I thought was my property. I suggested that we consider this an exchange of property since the tree would be attractive from the square below, or that the Mairie agree to sell the sliver of land at the end of the garden to me and not force me to relocate my wall. In the end the wall stayed where it was, on communal property.

My altercation with Gros over the location and valid existence of the right-of-way led to each of us getting our own surveyors. In my case I got a surveyor from Avignon, with the polish and looks of Giscard D'Estaing and the disdain that only a graduate of the best *Ecole Normale* can demonstrate. On a legal and intellectual level he demolished the village surveyor hired by Gros, but in the process understandably alienated the Mairie.

The fundamental source of our bad neighborly relations lay in the relative locations of our properties. With a garden built on the ruins of old houses, the water from irrigating my new garden terraces inevitably seeped down next to the wall of Gros's house. He claimed to be an architect, yet made no provision for drainage to protect his house against water from irrigation in my garden. He was later sued for the same defect in a house he restored in the village. Our arguments about the water and right-of-way resounded around our stone village with an agitated Mayor crying, *"Voyons, on est voisins, non?"* (Come on, you're neighbors, aren't you?). Gros's successor was no great improvement. He was a Swiss from Zurich, a town that one of my friends claimed from personal experience produced unpleasant neighbors. He lived up to that reputation by pouring cement over one of my sprinkler heads.

## La Mairie

The mayor's office and the school for small children adjoined our house, and we delighted in watching the children playing in the street. We were sad when they moved the school to much more adequate quarters down on *la plaine* (the flat).

One of my first visits in the village was to Mayor Ozias. My French friends said that we were lucky to have a communist rather than a socialist mayor, that he was likely to be more honest. Knowing that Ozias had recently returned from Moscow on a trip presumably funded by the communist party, I asked him his reaction. "Oh, it was all right," he said, "but I couldn't find a *Pastis* (the favorite alcoholic beverage in Provence) anywhere." The Mayor's real passion was hunting for truffles, often on the land of his constituents. He explained that pigs have the best noses for truffles, but trufflers prefer dogs because they do not eat the truffles. I suspected that the real powerhouse in the mayor's entourage was his tough, committed communist female secretary.

## Renting Saumane

French friends advised us strongly not to rent our house to anyone French. Under the law of 1948 established at a time of an acute housing shortage, it was virtually impossible to throw anyone French out of one's house in the winter months between September and March. It was even more difficult if children were involved. But we needed to rent and were impressed by Serge Majal, a physiotherapist, and his admittedly pregnant wife. The Majals left when asked, and their daughter Marie Prune was born shortly afterwards. To announce the birth, Serge Majal took a stunning Vermeer-like photograph of his wife's pregnant stomach.

## Our French Friends

Upon our arrival in Saumane, the Trottiers had introduced us to Jacques Reboul, an international green grocer, and Jeanette who ran a pharmacy, a prestigious position in small French towns. We formed what we called *le cercle des amis de Reboul*. Jacques was a generous bon vivant, whose figure showed that he enjoyed *la bonne chair*—a tribute also to the wonderful cuisine of Jeanette. Jacques complained about the small portions at an expensive nouvelle cuisine restaurant we took him to in Isle sur la Sorgue. Reboul's father had been mayor of Isle sur la Sorgue when the Germans threatened to shoot one hundred local citizens in retaliation for the killing of a German soldier. The mayor offered himself and his three sons as hostages, commemorated today in La Rue des 4 Otages. Politically Jacques was a quite conservative nationalist, an Anglophobe, and admirer of Pétain and Vichy France.

Alain Desmarez and his lovely mannequin wife Beatrice were the friends who made beautiful fake antique furniture for us and for others. It was Alain who had found our mason, Jean Paul. They specialized in restoring and selling old houses: first the beautiful farmhouse where the Abbé de Sade kept his library, la Vignerme; then an Italianesque house in Saint Didier. They had restored a house nearer to Mont Ventoux, Roman style, when their marriage broke up, leaving Beatrice to run their antique store in Isle sur la Sorgue. I was so impressed by the vitality of Beatrice's 90-year-old mother that I could not resist asking her recipe for youth. "My dear sir," she said, "I live on the sixth floor of an apartment in Paris without an elevator." Moved in with her children on the first floor, she quickly deteriorated.

Our oldest and closest friends were Collette and Phillipe Dreyfus from Paris and near us, in Lourmarin. Phillipe had been a paratrooper with the Free French in World War II. Collette had wanted to join him in England but the distinguished *préfet*, Jean Moulin, the head of France's most active noncommunist resistance movement, persuaded her that she could do more for France by working for him. He set her up in an art studio in Nice, which prepared false documents and helped Allied fliers to get out of France. Collette claimed to know who betrayed Moulin to the Nazis. Tortured in prison, he committed suicide to avoid betraying his network. But the scission between the pro- and anti-Vichy factions in France remained so explosive that Collette did not want to light another fuse.

Philippe was a great horseman (he used to ride for stag in northern France with Alain Desmarez's father). It was natural therefore that General Redgrave, cousin of Vanessa and a former commander of British forces in Hong Kong and of the British sector in Berlin, should turn to his good friend Philippe when he was suddenly named officer in charge of the Queen's Horse Guard. Redgrave admitted that he was not comfortable with horses, so Philippe took him out to the forest of Marly for riding lessons. Philippe was mortified when on the Queen's birthday his pupil Redgrave lost control of his horse. The old horse who knew the routine better than its rider had died the night before, and Redgrave was on a more excitable horse. The Queen was reportedly delighted by this breach of boring perfection.

It was thanks to the Dreyfuses that we found ourselves one evening guests of Mr. Marks of Britain's Marks & Spencer dining at Maxim's in Paris. Suddenly a pall fell over the dining room, and I could see a photographer positively sweating because he wanted to but dared not take a photograph of the wife of a British lord, whose short tutu skirt had caught on the back of her chair exposing her bare bottom. Formerly on the stage, she seemed to enjoy the attention.

And speaking of bare behinds, I'm reminded of a canoe trip down the Ardeche River ninety minutes from Saumane. French friends from Moscow days arranged for us and close Washington friends to take three canoes on the river. Warned about the heat, we were dressed like English explorers of the nineteenth century. I almost upset the canoe when asked the time by a statuesque, nude brunette when we were about to enter some whitewater. In fact, everyone during our eight hours on the river was in the buff except for us and an elderly gentleman who was fishing on a high bluff wearing only a Panama hat.

Also through the Dreyfuses we met Max and Liane de Gasquet, who lived on a beautiful farm near Boux, high above Apt, which had been a Protestant fortress during the wars of religion. Max was a rugged man of nature who used to go off riding in the mountains for days with his children, while Liane remained at home tending her garden and exercising her considerable talent as a painter. She died at a family home high above Aix en Provence, where she could watch the ever-changing mood and colors of Cezanne's Mont Sainte-Victoire.

Then there was Nicole de Bisset, widow and daughter of a noble French family, trying on a modest income to keep up the old family chateau at Lauriol near Carpentras. Hers was a strange and disturbing story. In her largely socialist region she was elected as a conservative to be mayor of Lauriol, presumably on the assumption that this little old widow would be a pushover. But when Nicole refused to authorize traditional *pots de vins* (pork projects) and uncovered a suspicious money-laundering operation, she received threats and an incendiary device was exploded outside her bedroom window. The *préfet* of Avignon, an old friend, refused to investigate, telling Nicole that he had received orders to this effect from the Ministry of Interior in Paris (which appoints the *préfets*). Nicole and her Conservative friends believed Mitterrand capable of anything, including, they told me, the murder of overly inquisitive members of his cabinet.

Among our other friends, Charles and Marie France Blagdon were an Anglo-French couple with three delicious daughters. He was in the wine- exporting business. We met them first when they lived in Saumane before they moved to Velleron. When they visited us in our cabin in Jackson Hole, the daughter Florence awoke at 5 a.m. her first day to see a huge moose looking at her on the screened porch.

Still living in Saumane, Pierre and Alice Daragnes were true Provence villagers. This stocky little self-made Renaissance man, who had built his own stone house, was a fine photographer, a talented painter of Provençal landscapes, and a collector of Roman coins that he had uncovered in our village and the hills with his magnetometer. More important for Saumane, he became a friend of the direct descendent of the Marquis de Sade through his research into the history of Saumane, for Pierre was also our local historian.

All these memories made it so hard for us to sell our home in Saumane, more difficult for me than selling either the ranch in Wyoming or our house in Maryland "on the hill." But being an absentee landlord with unexpected huge plumbing and gardening bills and incompetent housekeepers all argued for reducing our responsibilities. As Nan used to say, *"Il faut tourner la page,"* and turn the page we did.

We were pleased to be able to sell the house to a French couple who worked in the area and had a son working in Moscow—that old Russian connection.

## Our Sailing Adventures

Before leaving the subject of Saumane, I must tell you that we were not a sedentary family there. We would spend two or three weeks each summer serving as crew on a beautiful French ketch, *Katy II*, owned by a fanatic Foreign Service sailor Alan Logan.

On my first trip I sailed through Gibraltar to the Balearics and to Logan's home near Toulon. It was a rude initiation. Everything went wrong on the maiden voyage of this secondhand boat. The automatic pilot quit on me in the early morning watch, the slides on the mainsail jammed, and with a dead motor and under sail at night, we managed in rough weather to maneuver into a small Spanish port where a Scottish engineer fixed our oil line. I would watch anxiously as my captain and his half brother, neither of whom wore safety belts, would vanish under the spray trying to fix an ailing jib, wondering if I was going to be left alone as a novice to handle the boat. What kept up my morale was the sight of dolphins like silver torpedoes diving under the boat and of the shapely French-Laotian wife of Alan's half brother, who wore only a G-string.

In the Adriatic off the Bay of Kotor we thought we might capsize as a Bora Bora wind hit us broadside driving us toward the rocks. In Turkey we met *Katy II* near Troy, sailed through the Dardanelles into the Black Sea up to Varna, Bulgaria. Since the Bulgarian authorities would not let us sail at night to get to Constanta, Romania, in daylight, we took the train up to Bucharest to visit our old Moscow colleagues, Ambassador Roger Kirk and wife Betty and to revisit the wooden Russian Orthodox churches near the Moldavian border, with their amazingly preserved frescoes on the outside dating from the fifteenth century.

Our most beautiful sail was along the southern Turkish coast from Alanya, with its UNESCO-rated museum, almost to Bodrum. The solitary beauty of the little coves with their Greek ruins and mountains behind had an integrity that I somehow found lacking in the reconstruction of Knossos on Crete by Sir Arthur Evans, as beautiful as it was.

Our last and most exciting sail on *Katy II* was from Saint Petersburg, Russia, down to Klaipeda, Lithuania. We feigned sickness in the rough weather to enable us to spend our first night on an island that the naval authorities in Saint Petersburg had said was no longer a closed area, but about which the Russian Coast Guard had a different view. A large black Russian submarine surfaced to watch us from across the quay. In Liepau, Latvia, we saw the sad, rusting remains of the proud Soviet Baltic fleet, with hungry Russian sailors trying to sell everything from knives to classified charts of the Baltic. Off Tallinn it cost us $1,500 to get a Finnish dredge to cut us loose from a heavy fishing line around our propeller.

Our son Campbell persuaded us that a much pleasanter way to sail was in the Caribbean on a monohull, or catamaran, with a captain doing the work and a good cook in the galley. Our crewing days were over.

## Our Golden Years

With their usual panache the French have names for growing old: *troisième age* (third age), *croulant* (old fogey), *son et lumière* (sound and light, nearing the end of the road). As our old friend Chalmers Roberts wrote in his autobiography, "How did I get here so soon?" We don't know either, but here we are, determined to make the most of our remaining years.

The finest therapy for rolling stones like us is to just keep on traveling, satisfying our curiosity about places and cultures we have never visited. Gardens have been my passion since I used as a child to develop irrigation channels in our garden in the south of France to water the roses. Little wonder, therefore, that gardens have been a feature of our travels in retirement: the fantastic Van Dusen gardens in Vancouver, the rustic dacha gardens of Russia, the beautiful estate gardens of southern England contrived to appear so natural, the formal gardens of France and Italy, an exquisite garden built by a Japanese businessman outside Matsui, Japan, and, of course, the meditative beauty of the most famous Zen garden in Kyoto, with its swept pebbles punctuated by a few rocks.

I have been more interested in my Scottish than my English roots. And so I jumped at our son's suggestion in 2007 that we spend Christmas in Edinburgh, trying out kilts and single malts. Once again we were surprised by the quality of Scottish cuisine. Even haggis, properly prepared, can be tasty. In a little remote inn on an earlier trip we found a chef who spent two days each year sharpening his culinary skills in France. It was at this inn that I was reprimanded at breakfast by a lady guest for referring to my "Scotch" ancestors. I will never forget her saying: " If you are a Buchanan, Sir, Scotch is what you drink, Sir, and a Scot is what you are, Sir." Having held the moral high ground over my wife, a Campbell, who sided with England in the war with Scotland, I was embarrassed to see in Inverary Castle, the home of the Duke of Argyle and headquarters of the Campbells, that the Thompsons were a sept, or subclan, of the Campbells.

Selling our house in Maryland before the collapse of the real estate market has helped us indulge our love of travel. Taking advantage of their relatively cheaper prices, good accommodations, and normally superb lecturers, we have continued to sign up for tours with Elderhostel (renamed Road Scholar). In 1995 we took our first Elderhostel tour, to Thailand. We had planned to combine the tour with a visit to see the Angkor temple complex in Cambodia, but canceled our trip to Cambodia when an American lady archeologist was murdered en route to one of the temples. Instead we visited Burma, or Myanmar, with its own spectacular temple complex founded by the rulers of Pagon around the tenth century.

En route to Indonesia in 1996, via Seoul and Cambodia, provided the opportunity to visit the great complex of Angkor Wat. I concluded in discussion with our guide that probably the American archeologist had been killed because she rudely rejected demands that she pay for a security guard to take her to Banti Shrai, perhaps the most beautifully carved temple in the complex. We gladly paid $75 to have three young men on motorbikes drive ahead of us with their rusty Kalashnikovs. Our guide in Phnom Penh had lost all of her family to the Khmer Rouge, had watched them kill her uncle, and was forcibly married to a stranger who turned out to be a fine husband.

In Indonesia, our American guide, who had come to study Javanese dance, had married the leading prince of Bali. She described the linguistic and social mores in a royal family, where she was a commoner, where the language was high or low Balinese, depending on whom one addressed, and where the traditional greeting in the morning to her husband was; "Has Your Highness had his bath today?" On Java, the main island, we were awed by the Hindu temple complex of Borobudur, which predates Angkor by three hundred years. And on Sulawesi, there was the curious culture focused on the dead.

In 1999, a tour of the Sacred Ganges and the Himalayan Kingdoms, followed the usual route of the Taj Mahal and the cremation of the dead in Varanasi (old Benares), but then we took a train up to Darjeeling, a seventeen-hour trip that took fifty-two hours on a cockroach-and-rat- infested train, because of flooding from the Monsoon, We watched the sun rise on Kachenjunga, then on to Sikkim and the high point of the trip, Bhutan, with its own princely traditions, its houses decorated with a penis for fertility and good luck, its striking ochre-and-white-striped fortresses and its cleanliness, such a contrast to all its neighbors. The view flying past Everest on the way to Nepal was breathtaking.

We had tickets to fly to Tibet, but an English doctor in Kathmandu strongly advised against traveling at 12,000 feet, given our age and heavy colds. Instead we traveled to a game park to watch black rhinos from the back of an elephant.

As a poor substitute for the real Tibet, on our fortieth wedding anniversary in 1985 at age 61 I had joined a French youth group, Nouvelles Frontières, exploring the impoverished *gompas* (Buddhist temples) in the 14,000 foot Zanscar Valley near Ladakh, in preparation for climbing over the 17,500 foot Umasilla Glacier. The monks who served as porters on the glacier, with their plastic thong sandals, looked to the Dalai Llama as their spiritual leader. With the Kashmiri paying as little as possible to feed us, and given the terrain, I lost 25 pounds on that trip, but would not have missed it for anything. Nan, fearful of heights, stayed in Provence, eating far better than her husband.

Our next great adventure, in 2004, was an Elderhostel tour of Chile and a trip on our own to Peru and Bolivia. Easter Island, a five-hour flight from Chile, was the *pièce de résistance* of the Elderhostel tranche. Our guide on Easter Island was the son of the Chilean opposition politician Orlando Letelier murdered by Pinochet's hirelings in Washington. He showed us how the Micronesian immigrants destroyed themselves and a civilization best remembered for the great stone heads of Easter Island by destroying their environment.

In Peru, we learned that Machu Picchu is not the religious center it was first thought to be. It is rather an 8,000-foot-high winter resort for the upper class inhabitants of 11,000-foot-high Cusco. It remains an unforgettable, indeed spiritual experience.

Of the several Elderhostel tours we have made in Europe—to Italy and Eastern Europe—the most memorable was certainly the one to Sicily, starting with Taormina, that my mother had visited and adored around 1904. Our amusing Sicilian guides brought alive the layers of civilization that have marked Sicily: Phoenician, Arab, Greek, Roman, and Norman. The post–World War II American military government's great unthinking contribution, with the help of our Mafia don Lucky Luciano, we were told, was to help reinstall the Mafia, which Mussolini had exiled, as mayors and power brokers in the towns of Sicily.

Our Elderhostel tour of Andalusia in Spain in 2007 was also a reminder of the waves of civilization that have swept over the Iberian Peninsula: Phoenicians, Merovingians, Romans, Arabs, all leaving their marks. But for me the most poignant aspect of Andalucia was the reminder to us of what the Inquisition destroyed: a Moorish civilization far more advanced than that of Western Europe. It had been the conveyer belt for Hellenic culture to the West and an example of tolerance toward both Jew and Christian.

To visit southern India as we did in 2008 is to experience not only a continent ravaged by successive invaders, but a civilization that is perhaps the most complex on the face of the globe. Since most Indian empires in the north did not penetrate southern India, the temple complexes we visited were among the finest examples of original Hindu architecture, which, in turn, had influenced the construction at Angkor in Cambodia.

Our Albino Hindu guide was a leading authority on the early Roman trading posts established in southern India at the time of Christ. The apostle, "doubting" Thomas, is supposed to have come to Chennai and suffered martyrdom there. Our guide had just returned from a trip to Cambodia at the request of the Indian Archeological Society to compare the Dravidian temples of southern India with the temples at Angkor. We ended our tour with a side trip on our own to Rajastan, living in a luxury hotel in the middle of Lake Pichola in Udaipur. I was intrigued to learn that the Maharajah of Udaipur was named Sisodia, a variation I would assume of my Hindu prince stepfather's name, Seesodia.

We were particularly disappointed to have to abort a trip to Vietnam on health grounds. We had heard so much about the beauty of the country and the friendliness of its people. But for me Vietnam will always symbolize the folly of allowing ideological paranoia to involve us in wars in societies we do not understand, in historic conflicts that we cannot resolve.

# 17

# Mossy Wisdom of This Rolling Stone

Which brings me back by a long circuitous route to diplomacy. One of the joys of retirement is the opportunity it provides for old colleagues or new friends interested in foreign affairs to reminisce, reinforce their prejudices, and of course pontificate: "If I had been in charge of Vietnam, policy . . ." Our fuzzy memories of the past are much less important than the lessons we can distill from our experience that might help future foreign policy practitioners to avoid some of the mistakes our generation made.

Every issue in foreign policy is in some degree *sui generis*, but nonetheless instructive. Much of the gratuitous advice below is simply condensed common sense.

## Dos and Don'ts of Diplomacy

For understandable reasons, the writer has drawn on his experience with Russia to illustrate what, in his experience, are some of the diplomatic rules of the road.

- The point of departure for any policy is a realistic assessment of what one hopes to achieve.

- In this respect, Russia is no different from most countries. We hope to resolve differences peacefully and constructively, find areas where cooperation is in our mutual interest, identify ways to influence our protagonists, and protect our national security interests.

- In the case of Russia, our greatest mistake has been to assume that Russia will eventually see the light and conform to our views of democracy and the rule of law. George Kennan's advice in 1951, at the height of the Cold War, for dealing with Russia applies to relations with most countries. He said:

  > Let us not hover nervously applying litmus papers daily to their political complexion to find out whether they answer our concept of "democratic." Give them time, let them be Russian, let them work out their internal problems in their own way.

- Our effort to jump-start democracy and a market economy in the early 1990s without concern for Russia's traditions and national psychology was a major mistake with serious consequences, including the resurgence of autocratic rule

- To press for NATO enlargement, in violation of what the Russians saw as an agreement linked to the unification of Germany, was an affront to a proud people, reviving their historic paranoia about encirclement. We hailed the "color revolutions" along Russia's borders as a victory for democracy. The Kremlin saw them as the replacement of friendly governments with governments that were a potential threat to Russian security. It is sad to say that the pro-American attitude of many Russians during the Communist era, precisely because we were the declared enemy of the Kremlin, has been replaced in post-Communist Russia with suspicion and hostility.

- Understanding your protagonist does not mean to agree with his policies and viewpoint or to disagree with official U.S. policy. But the best protection against error is to maintain an open mind and question whether existing policy is the most effective way to deal with a protagonist.

- Certainly in the case of Russia, and indeed of most countries, the best way to influence them, and reduce suspicion or paranoia, is through patient negotiation in good faith on issues of common concern.

- Even when we find the behavior of our protagonist repugnant, it is more effective to speak in private of the consequences for our relations of such behavior rather than invite a confrontation through official public criticism that will require an official rebuttal. At the same time, we can affect a country's behavior by mobilizing world opinion against the offending state through discreet use of the public media.

- Americans are much more inclined than many states, and certainly than Russia, to see sanctions as an effective form of international pressure. We forget that many countries, including Russia, have been the target of sanctions imposed, sometimes unilaterally, by the United States, and they discourage the use of sanctions that might be turned against them. If key players like China and Russia with veto power in the UN Security Council oppose sanctions against, for example, Sudan, Burma, or Iran, other ways should be found to bring pressure to bear. It is easy for the United States to call for sanctions against Iran. It is not a neighbor of Iran, as is Russia, with a long tradition of friendly and profitable relations and a large Muslim population that would be vulnerable to subversion by an angry Iran.

- It follows that administrations should not make threats about sanctions or any other coercive measures that they are unable, or lack the will, to implement. It is dangerous to be regarded as a "paper tiger."

- It is equally detrimental to our international standing to be seen as a subversive state that will try to change your regime if it does not like your politics. Our tendency to demonize states we do not like as "rogues" tends to be counterproductive. It complicates negotiations with such states when they are in our interest, and in the case of North Korea and Iran have provided a plausible rationale of self-defense by seeking security in a nuclear deterrent, even if it violates the Nonproliferation Treaty.

- There are still influential neocon and other activists who argue that the United States could and should do more to intervene in the domestic politics of competitor states. It is this sort of dangerous thinking that would have us try publicly to affect the outcome of the ongoing historic debate in Russia between conservative nationalists like Prime Minister Putin and the more Western-oriented President Medvedev.

- To expect even the most skilled CIA agent to manipulate this Byzantine Russian world of clan, class, and personal rivalries would be the height of hubris. The likely result would be to discredit "Westerners" even more than they are today, as "tools of American imperialism.

- Attitudes and institutions in Russia will evolve at their own pace. The best we can do is by our policies and attitudes to encourage cooperation and the integration of Russia into the world community, react firmly when Russia tries to bully its neighbors or our allies, and avoid actions certain to provoke a nationalist and paranoiac backlash like the incorporation of Georgia and Ukraine into NATO.

Both Russia and the United States are burdened by attitudes and traditions that affect their ability to understand both the outside world and their own national interest. For example:

- As a nation of lawyers, we place an inordinate faith in the external forms of elections, formal checks and balances, and the rule of law. We do not always appreciate that the decision-making process, particularly in less-developed countries, may be appropriate for a society at their stage of development. Historically in Africa and today in Afghanistan with their Great Assembly, or *Loya Jirga*, there has been a tradition of decision by consensus, achieved through patient dialogue. Any agreement reached by tribal leaders may be one of the lowest common denominator, but it may at least avert conflict.

- Imbued by the Protestant work ethic and our Horatio Alger view that anyone can make it in society if he or she will work hard enough, we do not appreciate the attraction that paternalistic, social democratic states have in most parts of the world. Asked to choose between the predictability and security of a paternalistic government and the liberty and uncertainties of a democratic market economy, a majority of people will vote for security. As one Russian said during the turbulent Yeltsin years of fermenting democracy: "You can't eat freedom." This is certainly the case for millions mired in poverty in the developing world.

- Because other people envy and try to emulate our material success does not mean that they want to be just like us. They hope to catch up with us materially, but in their own way, at their own pace.

Patience is a virtue in diplomacy, as it is in personal relations, but one that does not fit well with the American character. Our impatience is a factor that foreign powers need to take into account in dealing with us. Impatience can also push American leaders and Congress to make rash decisions they will come to regret.

It is accordingly vital that the American president be a person of deliberate temperament, with experience, appropriate cynicism, and a moral compass to find his or her way through a maze of conflicting interests and advice in crisis situations. President Kennedy showed these qualities during the Cuban missile crisis. President Bush failed this test in the case of Iraq, ignoring the advice of senior statesmen and regional specialists, preferring to listen to people with something personal to gain from an American attack on Iraq. His reported statement to French President Chirac that, in attacking Iraq, America would be fulfilling a biblical prophecy, destroying Israel's enemies Gog and Magog, also suggests that he may have allowed religious beliefs to affect his judgment.

Statesman must take care in any effort to modernize a traditional society. There are often unintended consequences. In the case of Iraq, we have removed a brutal dictator but in the process strengthened the influence of Al-Qaeda and Iran in the region. The efforts of the Communist Party in Afghanistan and the shah in Iran to introduce modern reforms provoked Mullah-led revolutions in both countries. Russia has also experienced a backlash against efforts to modernize their economy. One of the costs of the breakdown of traditional societies is often an increase in instability and corruption.

In every society there is a tendency to inflate dangers emanating from the "enemy," be they domestic or foreign. Some circles benefit politically and financially from creating an enemy. In other cases, the fears are genuine. Regardless of motivation, fear-mongering adversely impacts diplomatic efforts to resolve differences peacefully.

In worst-case situations, fear can lead to dangerous incidents and even war. And war is an admission of diplomatic failure, to be undertaken only in the event of a demonstrable threat to America and its national security interests, or as part of an agreed international action directed against some egregious threat to human life and the law of nations. Only under conditions of extreme danger, should America consider going to war unilaterally.

President George H. W. Bush was able to put together an impressive coalition to liberate Kuwait in 1991 for two reasons: he worked assiduously to bring other countries aboard, and the issue was clear aggression against a sovereign state. Many of our allies considered the 2003 attack on Iraq as unprovoked aggression, whereas they had sympathized with our attack on Afghanistan following 9/11. Both wars illustrated the danger of engaging in combat when we did not really understand our enemy and had not prepared an exit strategy.

As one might expect, we hear familiar voices warning us to beware of an emerging "Russian threat." These doomsayers choose to or genuinely misunderstand that Russia's sometimes aggressive rhetoric and posturing are designed to cover up the fundamental weakness of the Russian state.

- The Kremlin is trying to build up a strong, modern state on a foundation of sand. Its demographic crisis of low birth and high mortality rates, and a medical system that lacks the resources to deal with rampant alcoholism, AIDs, resistant TB, diabetes, and heart disease has serious long-term implications for its labor force and military preparedness.

- By 2050, when the American population will be approaching four hundred million, the Russian population may have shrunk to little more than one hundred million.

- The weakness of the Russian military was glaringly demonstrated in both the war in Chechnya and Georgia.

- Furthermore the Kremlin faces a serious problem of radical Muslim separatists in the Caucasus, centered on Chechnya, and of a large Muslim population in the rich republics of Bashkortostan and Tatarstan, which must be treated with kid gloves.

- Economically, Russia survived the 2008–2009 recession fairly well, but it is still dependent for much of its wealth on exports of oil, gas, and minerals. Despite all their talk about catching up technologically and industrially with the West, of establishing their own Silicon Valley, not much has happened.

- As for the fear that Russia will gradually regain control over its former empire, there are no signs that the states carved out of the empire are willing to give up their sovereignty and return to the Russian fold. In short, to speak of the threat of an expansive, imperialist Russia may be useful politically for right-wing politicians and members of the military- industrial complex, but it is quite unrealistic for the foreseeable future.

- Our leaders need to be fully aware of this reality; but to mock Russia's weakness, as Vice President Biden did, only serves to exacerbate the already substantial anti-American feelings in Russia.

There is much talk in both Russia and America of the need for better public diplomacy to combat negative images of both our countries. But public diplomacy cannot exist in a vacuum. In the final analysis, its success depends on confidence in the country and its message. Our public image is still too much that of a people hunkered down in their Green Zones, worried about the next terrorist attack. Public diplomacy will have something to work with once we regain our American self-confidence, leave our tortoise shells, and reach out to the people around us, focusing on the opportunities to improve their world and ours.

A central goal of U.S. foreign policy should be to restore that image of America as a Shining City on a Hill, a naïve society, perhaps, but one that can be trusted. It is the vitality of our institutions, our ability to speak openly of our flaws and address them, the demonstrated creativity of our population, and a visible self-confidence that verges on being cocky, that will reassure our allies and make nations that do not like us at least respect us. If we are careful to preserve a big stick, we can then speak softly, as we should, but we will still be heard.

It is of course fun to tell one another what each of us would do if he or she were king or queen for a day—to theorize like any academic without responsibility for one's actions and be free of pressing deadlines, of lobbyists peddling their wares, of important projects that have to be shelved for lack of money, insulated from the maddening crowd of pundits demanding action on the key issues of the day, not having to make the inevitable compromises without which controversial domestic or foreign legislation will never pass the Congress. We are so lucky to be retired.

In real life, as we recognize, policymakers would find it impossible to live up to the standards that experience has shown to be wise. Still, there is value in setting standards by which to measure the performance of our successors and, if we are frank, of ourselves. While the journey down memory lane can sometimes be a blow to the ego, it is also a reminder of how rich our lives have been in the Foreign Service.

## A Final Mossy Journey

When some of us say, "My traveling days are over," we forget a minor detail. The greatest journey of all lies ahead. I am frankly not excited by most of the brochures. The deluxe tour to the Christian heaven sounds a bit high priced and boring. The economy tour to Hell sounds more interesting, but I don't think I could stand the heat. To reach Buddhist Nirvana through a journey of meditation requires more discipline than I possess; meditation makes me sleepy. An eternal journey through space with trillions of other souls would be too crowded and claustrophobic for my taste. Reincarnation sounds like an interesting journey, but I'm told that my destination will depend on what sort of life I have led. In my case, I might return, deservedly, as a cockroach. I don't want to take that risk, if I have any say in the matter.

No, I am an earthbound creature who must look to the earth for guidance. It tells me that I am no different from the birds, bees, trees, and bushes, even mossy rocks and minerals, an inseparable part of an amazing cycle of nature. As Keats reminded us in his poem, "Four Seasons Fill the Measure of the Year," we all have our "lusty spring," our summers when "luxuriously [we] chew the honied cud of fair spring thoughts," our autumns when with folded wing we are content "to let fair things pass by unheeded as a threshold brook," and our winters of "Misfeature," reminding us of our "mortal nature." To write one's memoir is to relive all these stages of growth and decline.

To see one's self as part of an infinitely complex process of change and evolution of all things human and material is somehow reassuring. I cannot pretend to know what set this whole amazing world spinning on its axis, but I do know that, for me, this immutable, unfathomable cycle of nature is God Manifest. And in returning "dust to dust" I will be fulfilling my divine destiny.

# Index